Contents

How to use this book
SPM2.1	3D objects	
SPM2.1	Octahedrons	
SPM2.1	3D objects	5
SPM2.2	Angles	6–7
SPM2.3	Direction	8–9
SPM2.3	Angles	10–11
SPM2.4	Coordinates	12–15
SPM2.5a	Acute, obtuse and reflex	16–17
SPM2.5b	Angles	18–20
SPM2.5b	Angles in a triangle	21
SPM2.5c	Angles	22
SPM2.5c	Drawing angles	23
SPM2.6a	Parallel and perpendicular	24–25
SPM2.6a	Triangles	26–27
SPM2.6a	Polygons	28
SPM2.6a	Quadrilaterals	29
SPM2.6b	Circles, ellipses and semi-circles	30
SPM2.6b	Circles	31
SPM2.7a	3D objects	32–34
SPM2.7b	Objects with curved faces	35
SPM2.7b	Nets of cones and cylinders	36
SPM2.8	Combining shapes	37
SPM2.8	Tessellating	38
SPM2.8	Angles in tessellations	39
SPM2.9	Symmetry	40
SPM2.9	Reflections	41
SPM2.9	Rotations	42–43
SPM2.9	Patterns	44
SPM2.10	Polygons	45
SPM2.10	Triangles	46
SPM2.10	Quadrilaterals	47

How to use this book

Each page has a title telling you what it is about.

Instructions look like this. Always read these carefully before starting.

This shows how to set out your work.

These are Rocket activities. Ask your teacher if you need to do these questions.

Read this to check you understand what you have been learning on the page.

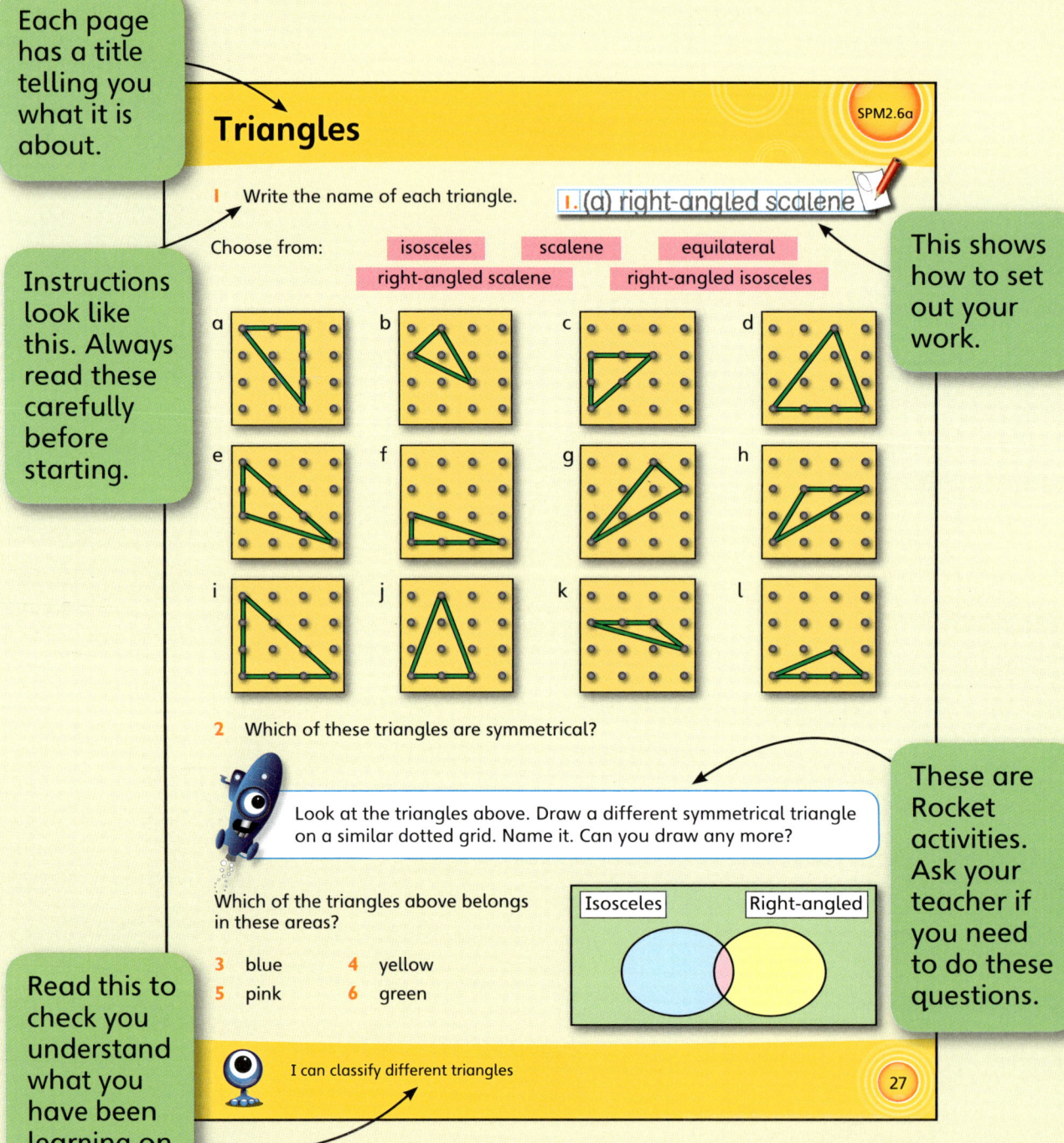

3D objects

Write the name of the object made by each net.

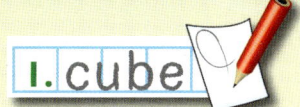
1. cube

1    2 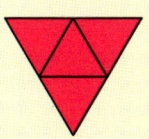   3

4 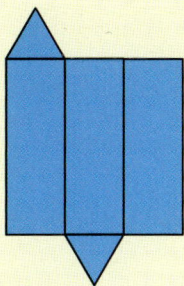   5    6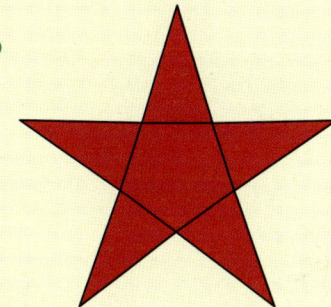

Which of these nets cannot be folded to make a cube?

7 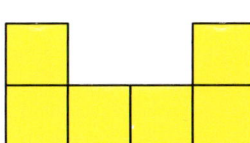   8 (image)   9

10 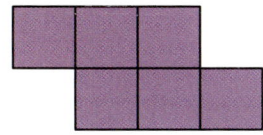   11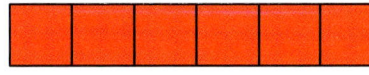

Check by copying them onto squared paper, cutting them out and folding.

I can look at a net and visualise the 3D object it makes

Octahedrons

Some of these nets fold to make an octahedron. Look at each shape and decide whether the net makes an octahedron. Write yes or no.

Use interlocking triangles to help you.

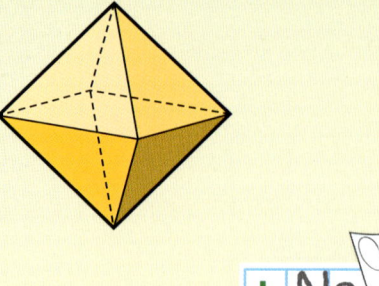

1. No

1
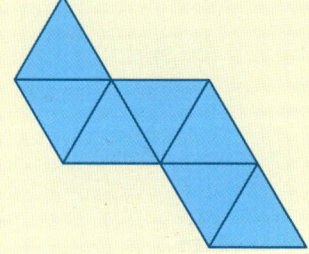

2
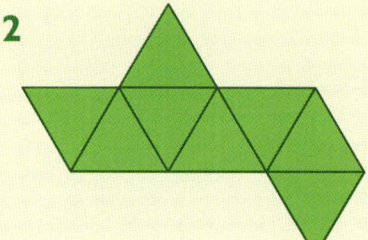

3
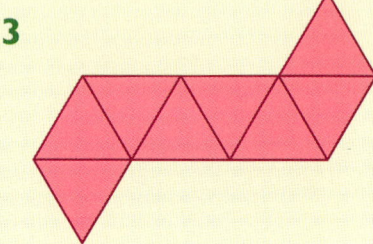

4

5

6
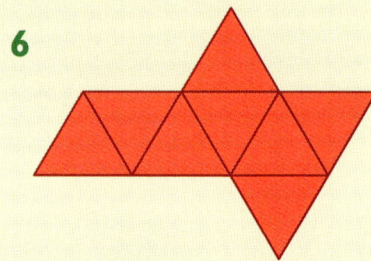

7
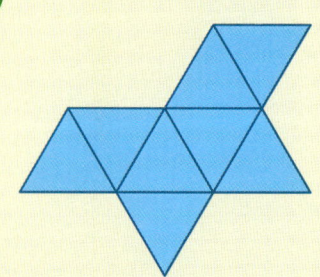

8

9
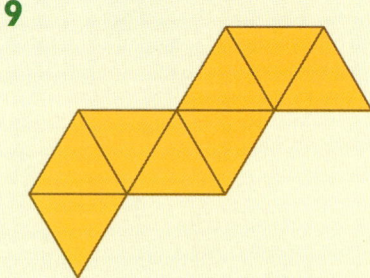

I can look at a net and visualise the 3D object it makes

3D objects

SPM2.1

Write the name of the object made by each net.
Write 'none' if it is impossible!

1. cube

1
2
3
4
5
6
7
8
9

Sketch a net that would make each of these 3D objects.

10
11
12

Choose something in the classroom and sketch its net.
Can your partner guess what object you chose?

I can visualise nets and the 3D objects they make

Angles

How many right angles are there in each shape?

1. 2 right angles

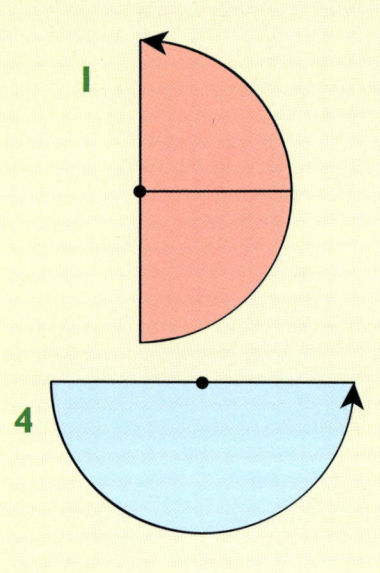
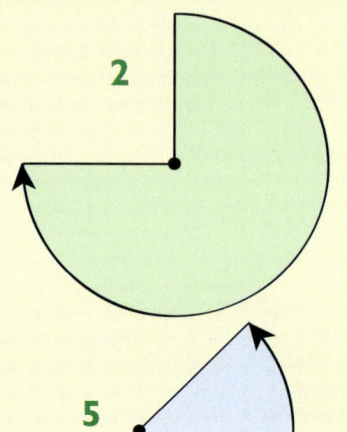
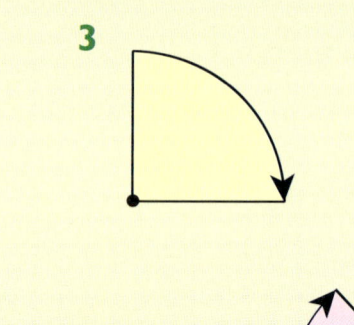

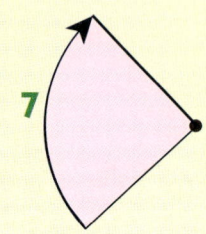

Now write the number of degrees in each angle.

2 right angles

Look at the questions below.
If the compass needle turns clockwise, how many right angles does it turn through?

8 starts pointing east, ends pointing west
9 starts north, ends south
10 starts south, ends east
11 starts east, ends south

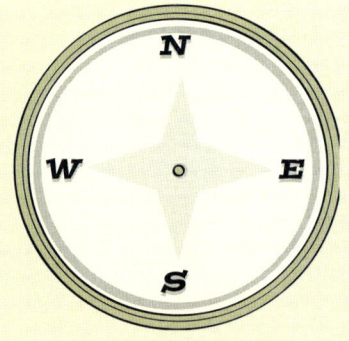

If the needle turns anticlockwise, how many right angles does it turn through each time?

Work with your partner. Draw a line on some squared paper. Then draw a line at right angles to it. You then draw a line at right angles to their line. Keep going to make a pattern.

I can recognise right angles and relate them to the four compass points

Angles

All these turns are clockwise. Write the direction of each plane after they turn.

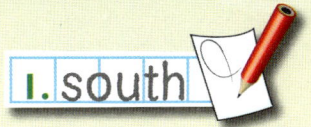

1. flying north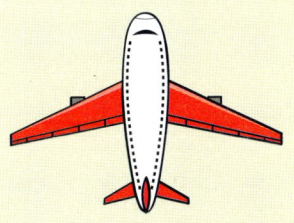
turns 180°

2. flying east
turns 90°

3. flying north-west
turns 90°

4. flying south-east
turns 180°

5. flying west
turns 90°

6. flying north-east
turns 90°

How many degrees are being turned?

7. 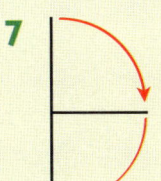 180°

8.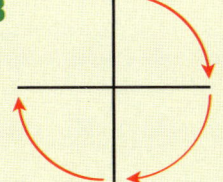

9.

10.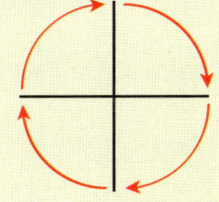

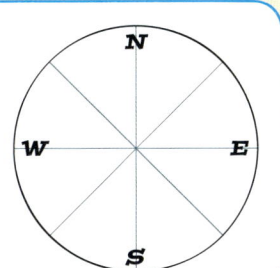

Start with a compass face. How many different ways can you shade a right angle turn?

I can recognise right angle turns and relate them to the four compass points

Direction

↑ is north. Write the direction of the arrows.

1. north

1 2 3 4

5 6 7 8

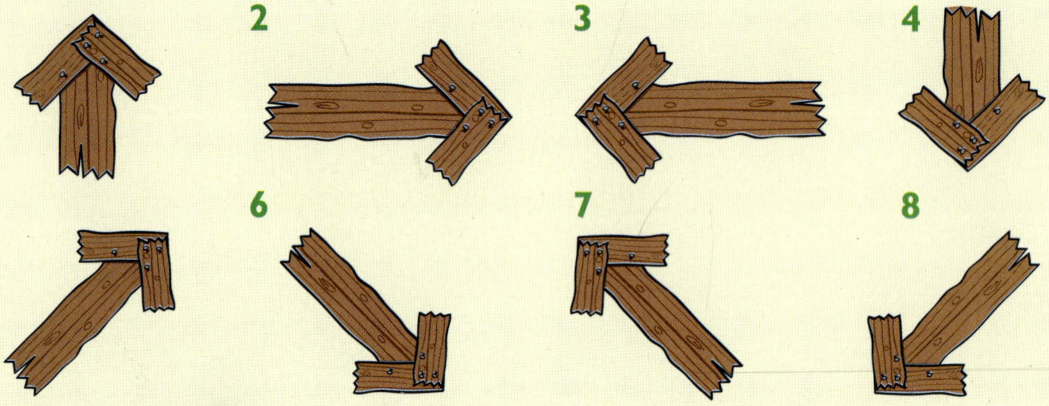

Write some pairs of directions that are opposite each other, e.g. north and south.

Where do you get to if you go:

9. Riptide

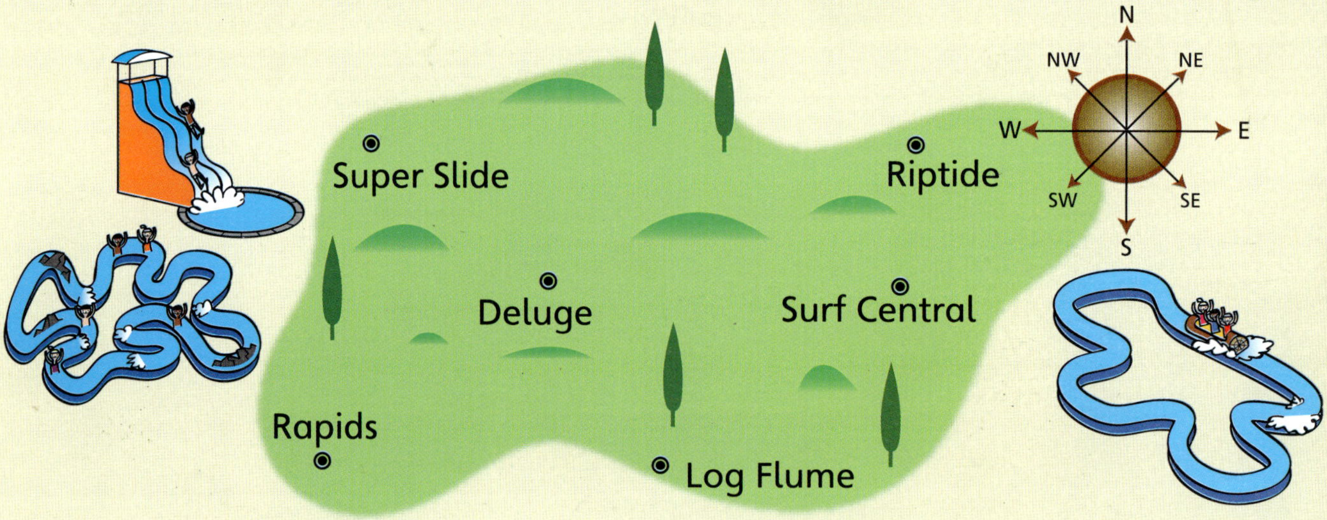

9 east from Super Slide?
10 south from Riptide?
11 west from Surf Central?
12 north from Rapids?
13 north-east from Log Flume?
14 south-west from Deluge?

I can make journeys involving compass directions

Direction

SPM2.3

Write the approximate direction to go from one city to the other.

1. south-east

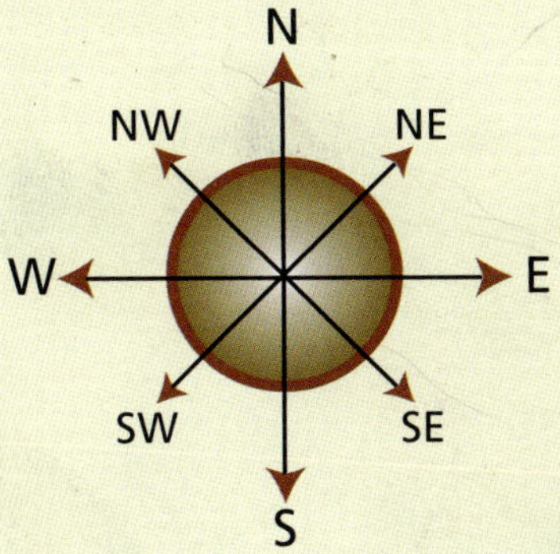

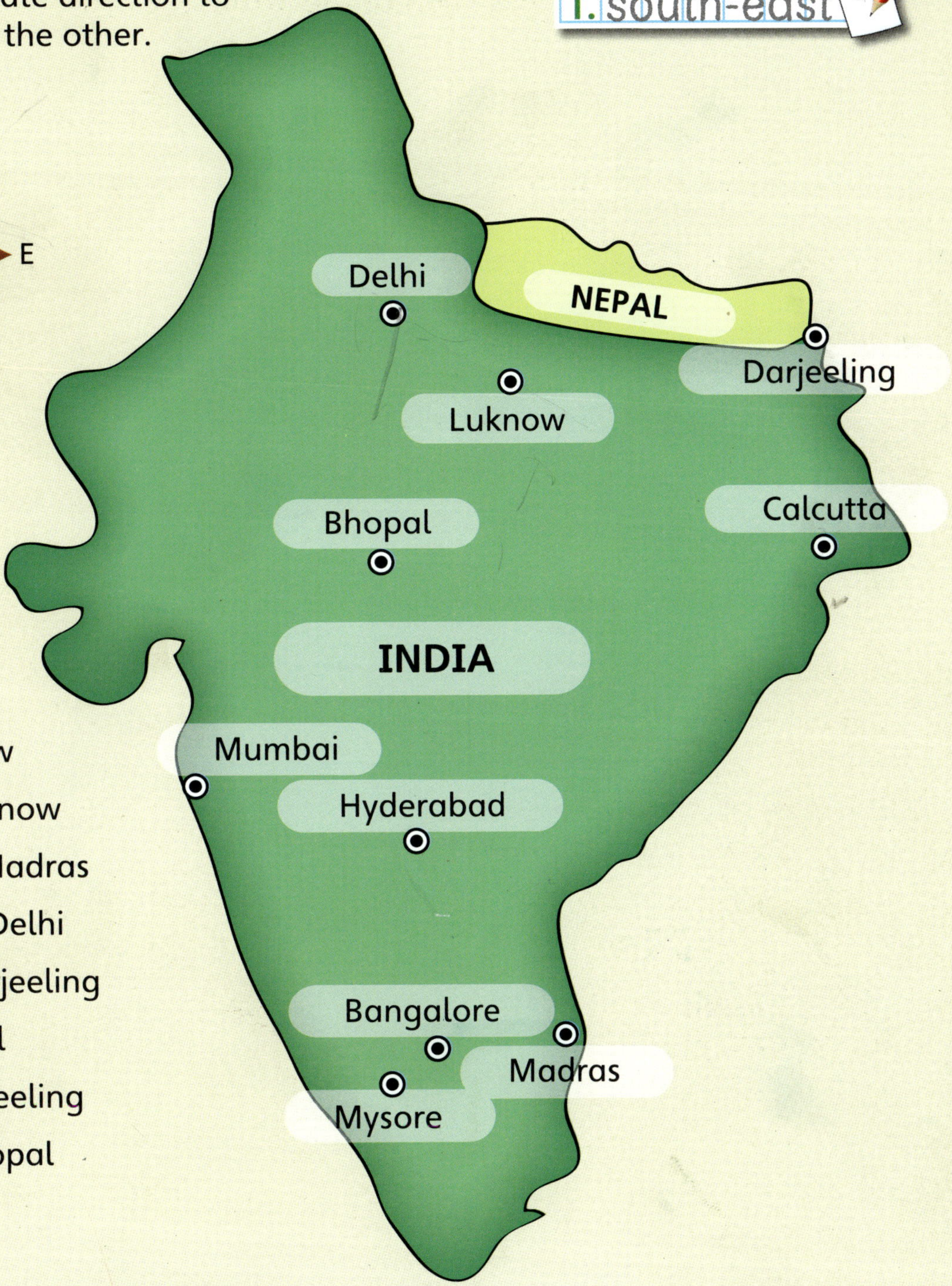

1. Delhi to Luknow
2. Mumbai to Luknow
3. Bangalore to Madras
4. Hyderabad to Delhi
5. Calcutta to Darjeeling
6. Delhi to Bhopal
7. Mysore to Darjeeling
8. Calcutta to Bhopal

Look for a map of India in an atlas. Write some direction questions of your own.

I can make journeys involving compass directions

Angles

Write the size of these angles in degrees.

1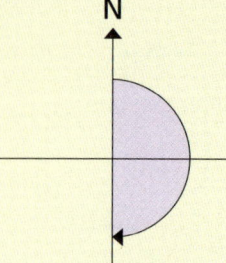
2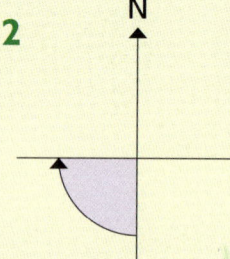
3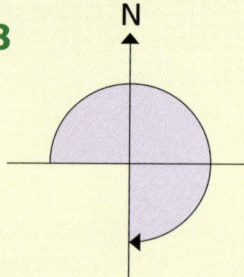
4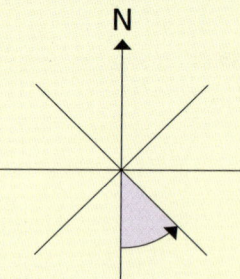

5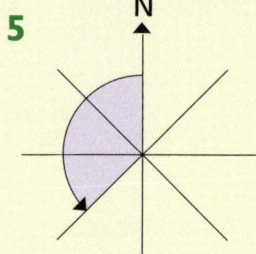
6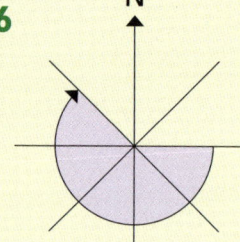
7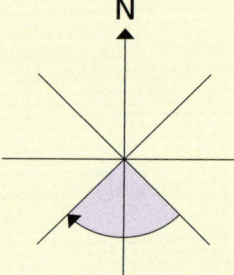
8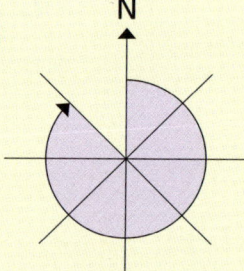

I face north.
Which direction will I face if I turn clockwise through:

9   135°?   10 360°?   11 180°?

12 45°?   13 315°?   14 270°?

15 450°?   16 225°?   17 405°?

What if I was facing south-west and made the same turns?

I can relate compass points to angles

Angles

Write the number of degrees in each turn.

1. 90°

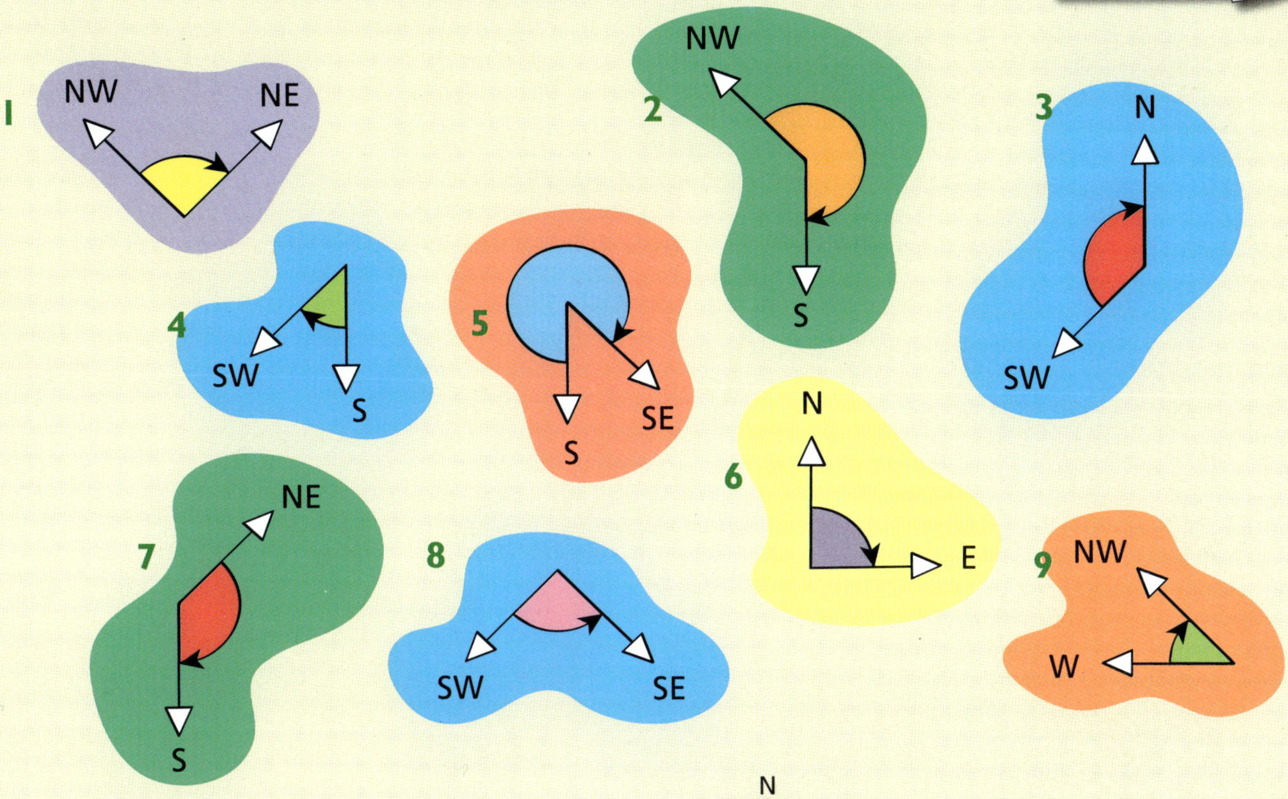

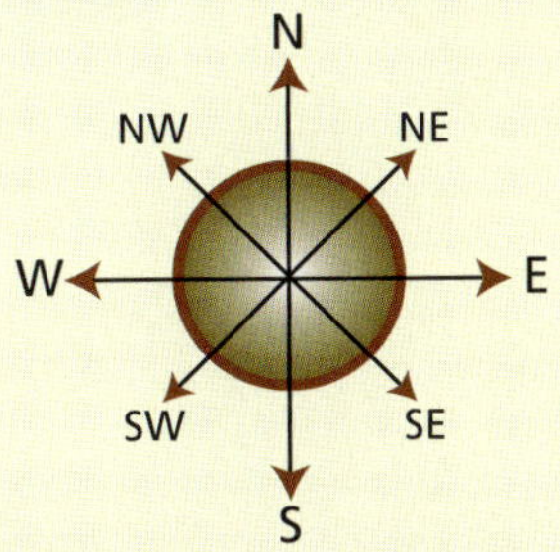

(Diagrams 1–9 showing compass turns)

1. NW to NE
2. NW to S
3. N (turn)
4. SW to S
5. S to SE
6. N to E
7. NE to S
8. SW to SE
9. NW to W

A weather vane turns through different angles. Write the angle it turns each time as it moves:

10. 45°

10 clockwise from N to NE
11 anticlockwise from S to E
12 clockwise from NE to S
13 anticlockwise from N to S
14 clockwise from N to SW
15 clockwise from NE to NW.

Write some more of your own for your partner to solve.

I can relate compass points to angles

Coordinates

1. Write the coordinates for each place.

   1. Bristol (7,2)

For which places on the map is the first coordinate more than the second coordinate?

I can give the coordinates of points on a map

Coordinates

Write the coordinates of the vertices of each shape.

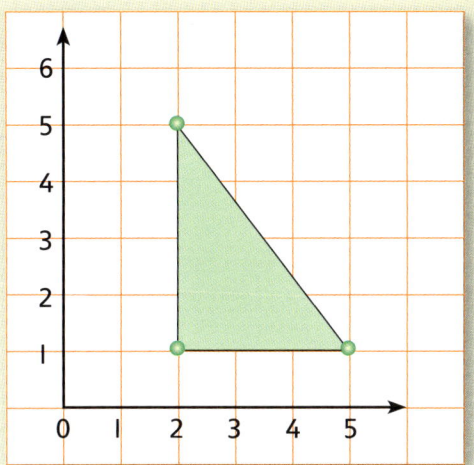

1

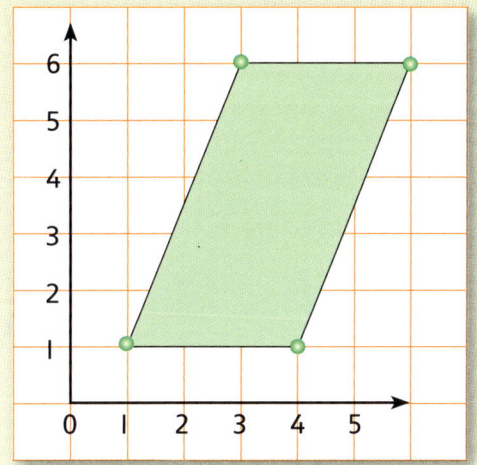

2

Draw 8 × 8 coordinate grids on squared paper. Plot the points below and join them to make a shape. Write the name of the shape.

3 (3, 5), (7, 5), (5, 2)

4 (2, 3), (2, 6), (6, 3), (6, 6)

5 (0, 2), (0, 8), (2, 4), (6, 4)

6 (1, 1), (5, 1), (1, 7), (5, 7), (8, 4)

Make up another problem like questions 3 to 6 for your partner to work out.

Write coordinates of a square with a missing point. Your partner says the missing coordinate. Repeat for a rectangle.

I can give the coordinates of the vertices of a shape drawn on a grid

Coordinates

Write the coordinates of: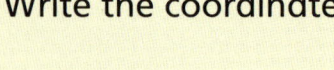

1. Forton
2. Elmbridge

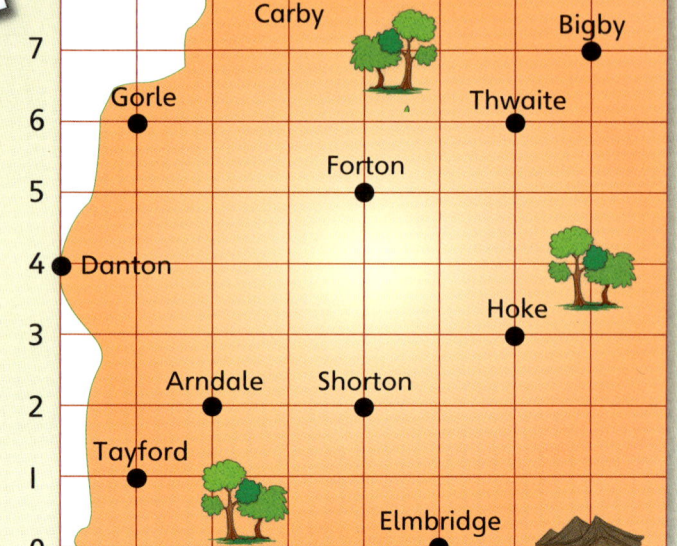

Write the horizontal coordinate of:

3. Carby
4. Gorle
5. Danton
6. Arndale

Write the vertical coordinate of:

7. Hoke
8. Bigby
9. Tayford
10. Shorton
11. Write the places where the horizontal coordinate is 4.
12. Write the places where the vertical coordinate is 6.

Write the place you reach if you start at:

13. (2, 5), go east two squares
14. (7, 4), go north three squares
15. (4, 3), go west three squares, north three squares
16. (5, 7), go south five squares, west three squares
17. (0, 2), go north-east four squares, south one square
18. (3, 0), go north-west two squares, north six squares, south-east three squares.

I can use coordinates to make journeys on a grid

Coordinates

1 Write the coordinates of each point.

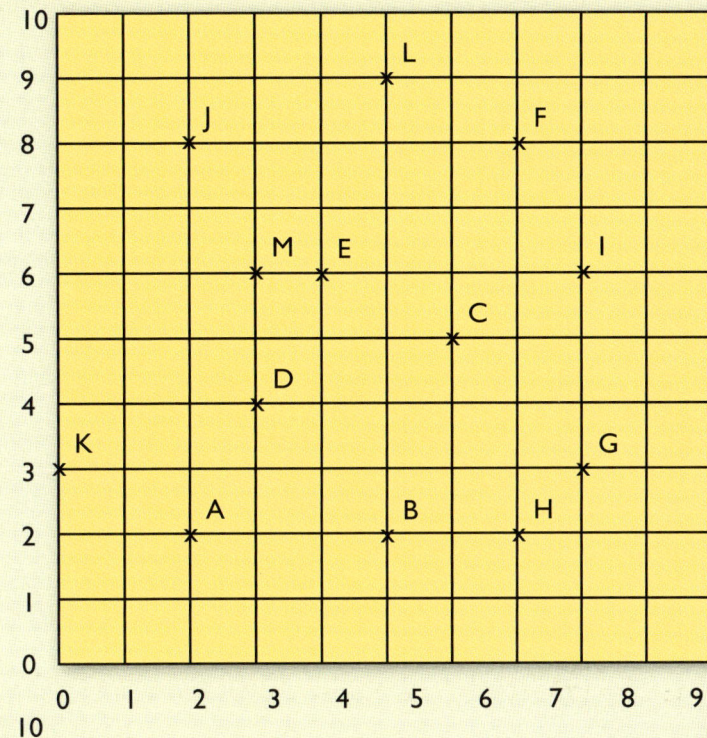

HINT Walk before you fly.

Write the distance between:

2 L and B   **3** A and J   **4** F and J   **5** B and H

Follow the instructions. Write the coordinates of where you land.

6 start at L, down 3   **7** start at C, up 3

8 start at A, left 2   **9** start at H, right 2

Write the coordinates of the point exactly half way between:

10 B and H   **11** F and H   **12** M and D

Make up a coordinates problem for your partner to solve.

I can use coordinates to solve problems

Acute, obtuse and reflex

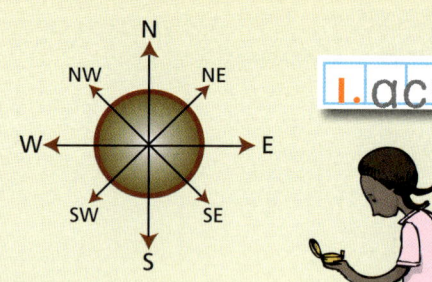

Sarah turns clockwise. What type of angle does she move if she goes from:

1. N to NE?
2. E to SW?
3. SE to S?
4. NW to NE?
5. SW to E?
6. N to NW?
7. NE to E?
8. N to SE?
9. SE to SW?

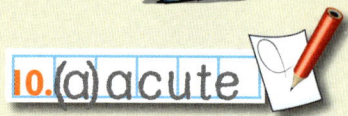

Write acute, obtuse, reflex or right angle each time.

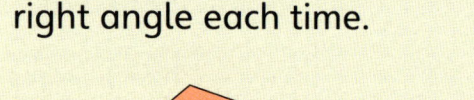

10.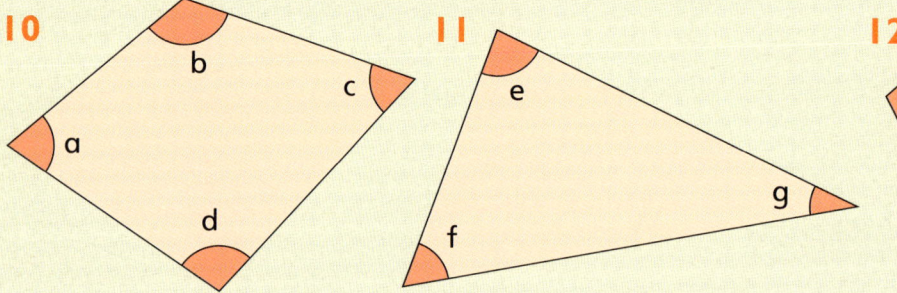

12.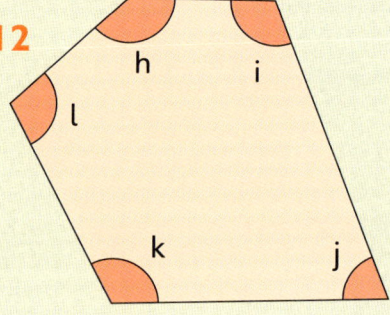

13.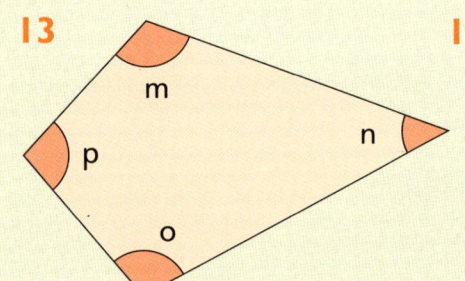

14.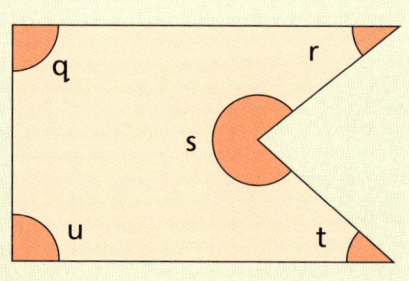

15.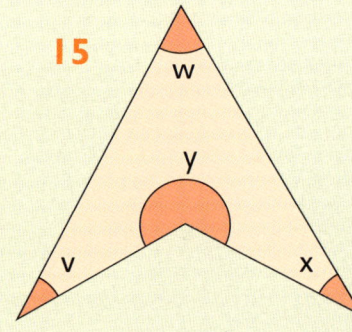

True or False?
The angles of an isosceles triangle are all acute angles.
The angles of a regular hexagon are all obtuse angles.
A triangle cannot have more than one obtuse angle.

I can recognise acute, obtuse, right and reflex angles

Acute, obtuse and reflex

Copy and complete the table to show the types of angle in each polygon.

Pinboard	Acute	Right angle	Obtuse	Reflex
1	2	1	0	0

1 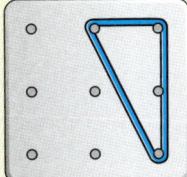 2 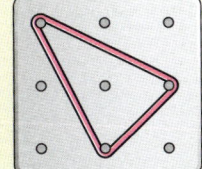 3 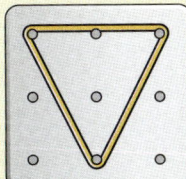 4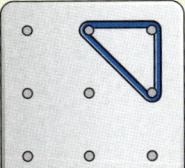

5 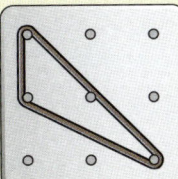 6 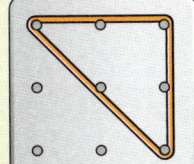 7 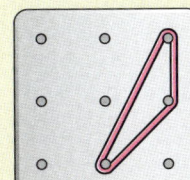 8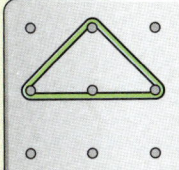

9  10 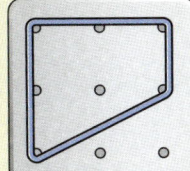 11 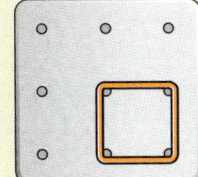 12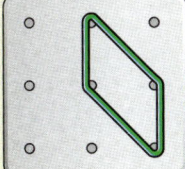

13 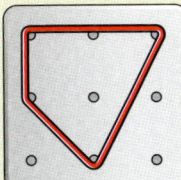 14 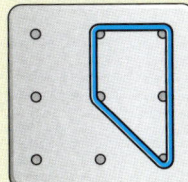 15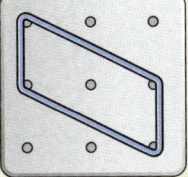

 Investigate the angles in different pentagons drawn on a 3 × 3 grid.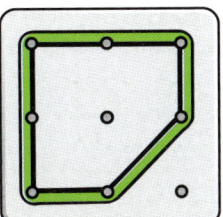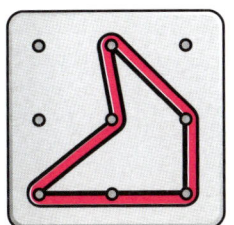

 I can identify acute, obtuse, right and reflex angles in shapes

Angles

Write the angles measured by these protractors.

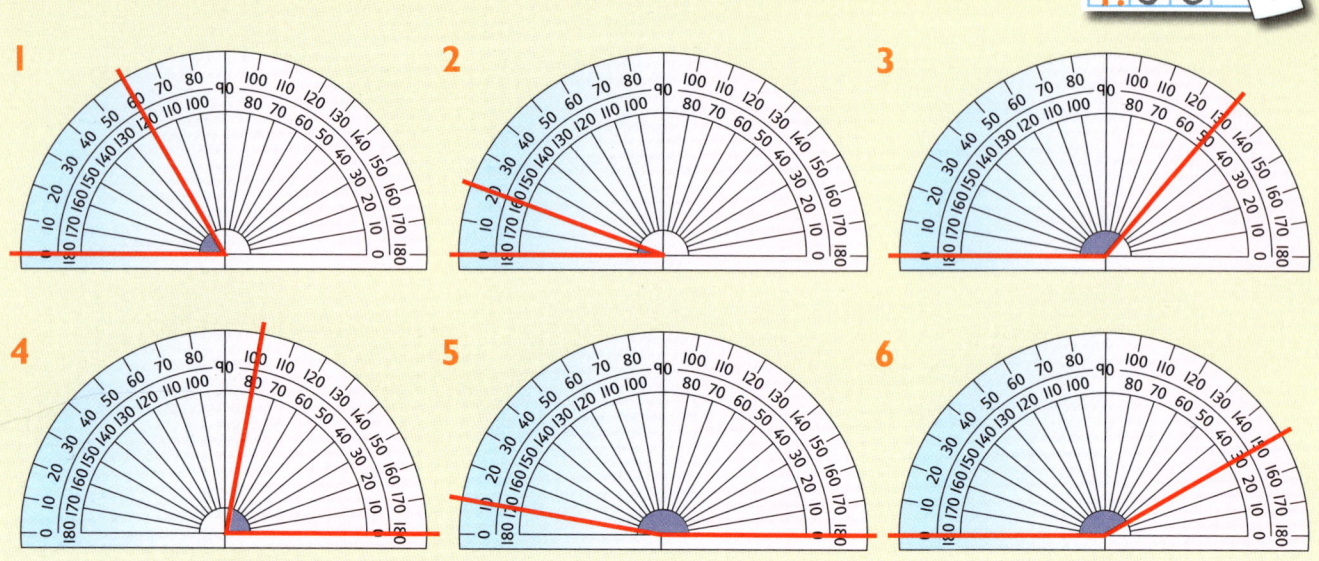

For each cake slice: (a) estimate the angle and (b) measure the angle. How good are your estimates?

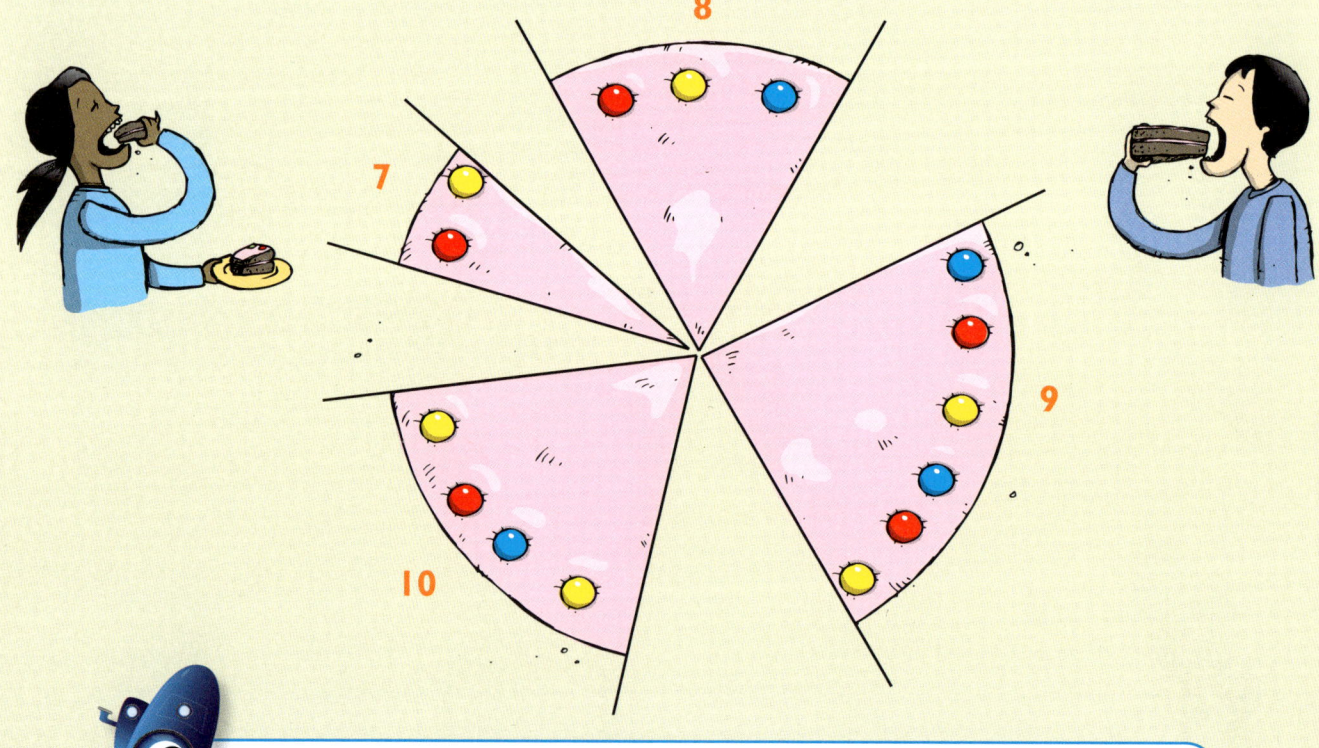

What is the angle of each slice if a cake has been cut into 3 equal slices? What about 4, 5, 6 ... equal slices?

I can estimate and measure angles with a protractor

Angles

Write the approximate angle shown on each protractor.

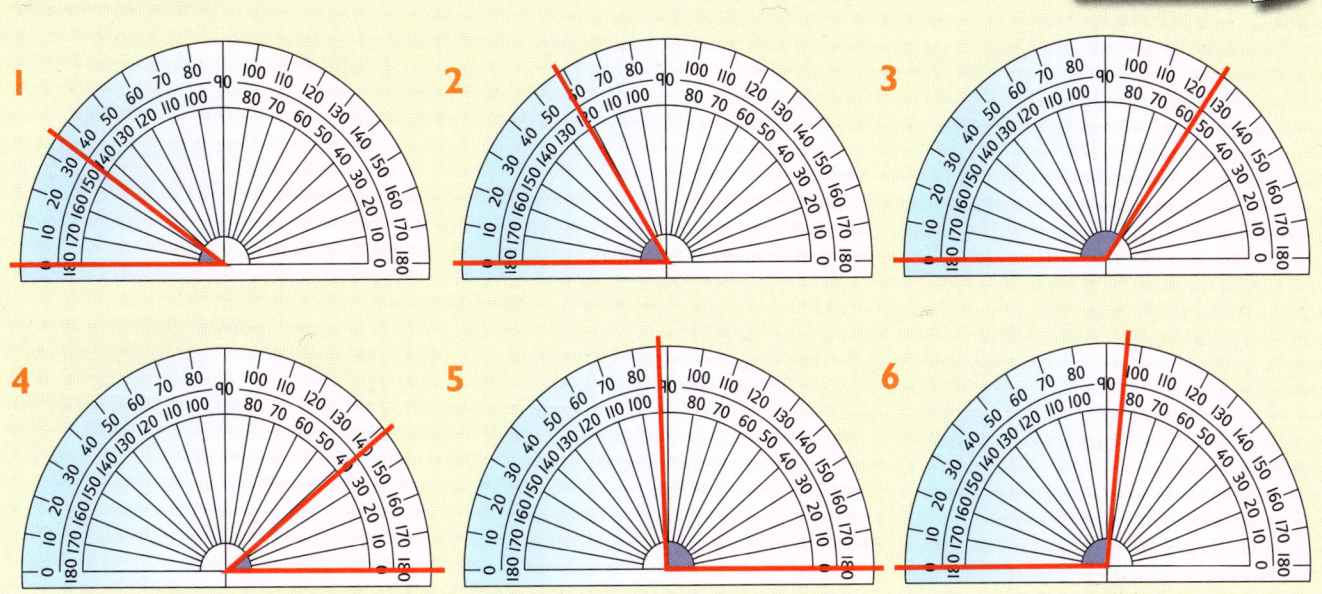

7   Estimate the size of each angle. Then measure them using a protractor.

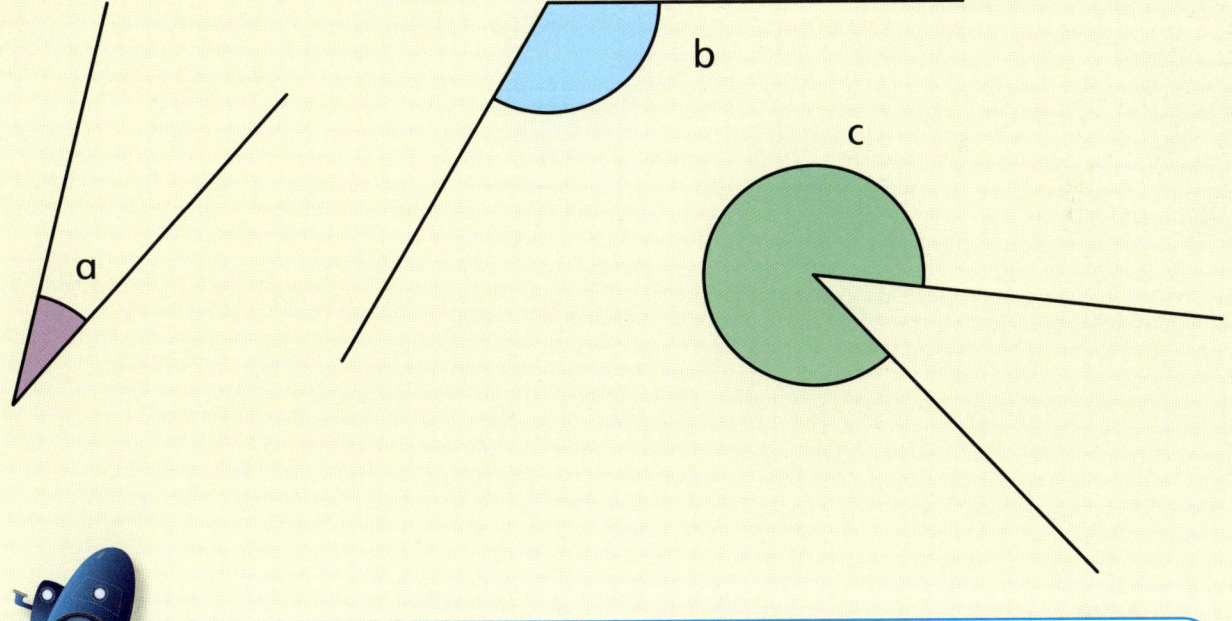

Draw an obtuse angle. Draw another angle that you estimate is half the size. Check your estimate by measuring both angles.

I can estimate and measure angles

Angles

Write the size of each angle.

1. 105°, ...

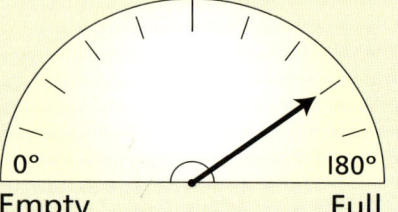

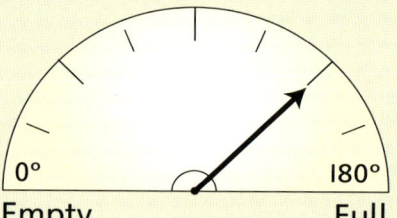

Use a large coordinate grid.

Draw a straight line from the origin (0, 0) to the point (6, 3).

Measure the angle made with the x-axis.

Find other points that make the same angle.

Repeat for other straight lines. What patterns do you notice?

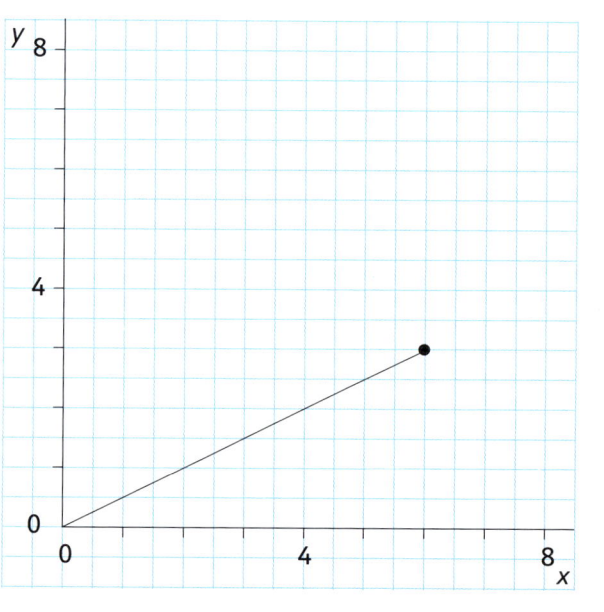

I can estimate, measure and work out angles

Angles in a triangle

SPM2.5b

Use a protractor to measure the size of each angle.

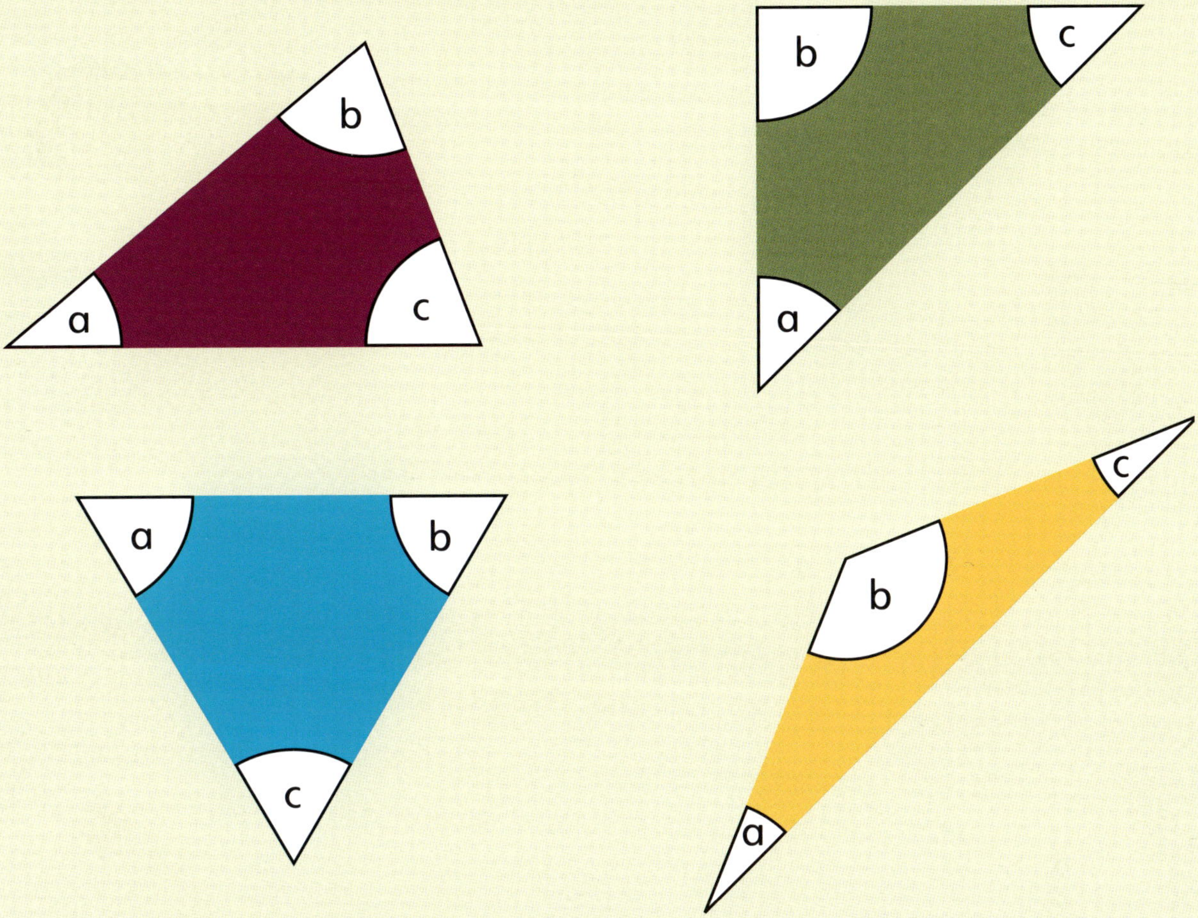

5 For each triangle, add the three angles together. What do you notice?

6 Draw three large quadrilaterals.

Measure accurately the four angles in each.

Find the total of the four angles.

What do you notice?

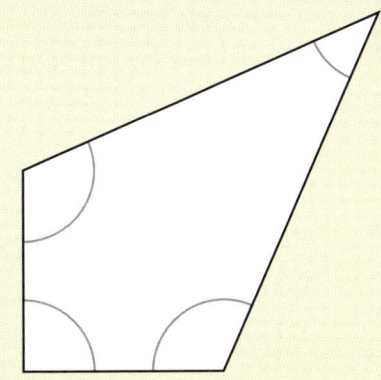

I can estimate, measure and work out angles in shapes

21

Angles

Use a protractor to draw these angles.

Acute angles: **1**   60°   **2**   75°   **3**   38°

Obtuse angles: **4**   110°   **5**   155°   **6**   127°

> Work with your partner. Both draw an angle you estimate is about 72°, no protractor allowed yet. Now measure each other's angles. How close were you?

Work with your partner. Each draw one of these angles. Cut out both angles carefully and fit them together, then say whether they make a straight line.

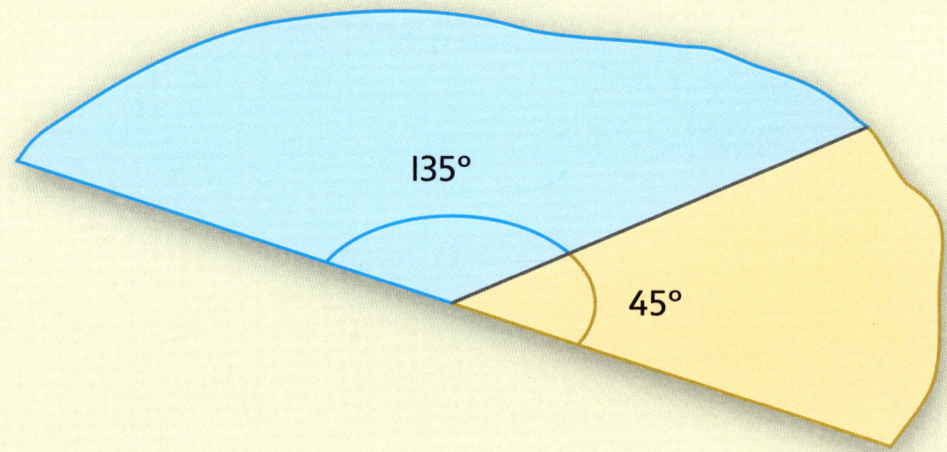

Now do the same thing with these pairs of angles.

8   38° and 142°   **9**   145° and 25°   **10**   107° and 63°

11   86° and 94°   **12**   161° and 19°

I can draw acute and obtuse angles

Drawing angles

Use a protractor to draw these angles.

Reflex angles:
1. 280°
2. 310°
3. 195°
4. 225°
5. 326°
6. 207°

Work with your partner. Both draw an angle you estimate is about 219°, no protractor allowed yet. Now measure each other's angles. How close were you? Now try it for 352°.

Work with your partner. Each draw one of these angles. Cut out both angles carefully and fit them together, then say whether they make a full turn.

7. 275° and 85°
8. 218° and 132°
9. 204° and 156°
10. 67° and 283°
11. 46° and 314°
12. 12° and 348°

I can draw acute, obtuse and reflex angles

Parallel and perpendicular

1. Look at each of the shapes below. Does the shape have parallel sides? If yes, write the number of pairs of parallel sides.

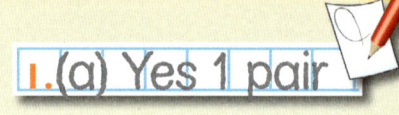

2. Does the shape have any pairs of perpendicular sides? Write yes or no.

a  b 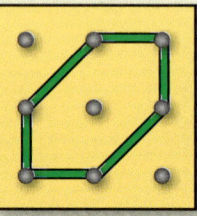 c 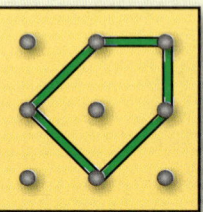 d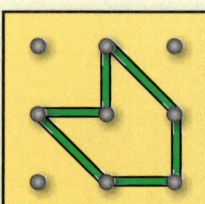

e 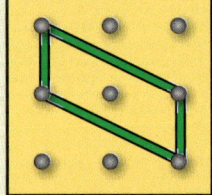 f 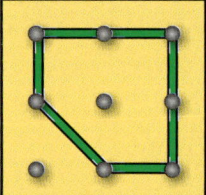 g 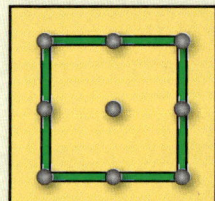 h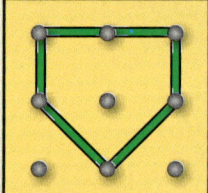

i 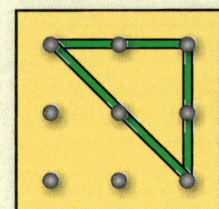 j 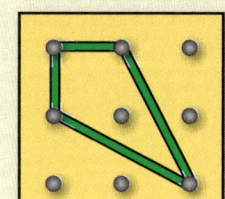 k 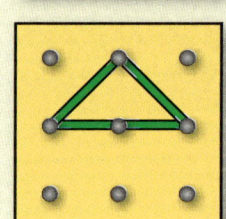 l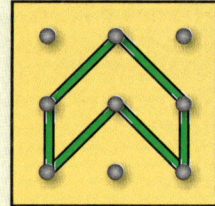

Draw 4 × 4 squares on squared paper.

Investigate how many different shapes you can draw that have:

- one pair of parallel sides
- two pairs of parallel sides
- three pairs of parallel sides.

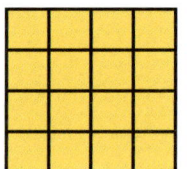

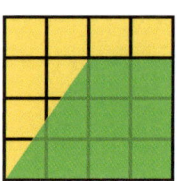 

I can identify parallel and perpendicular sides of shapes

Parallel and perpendicular

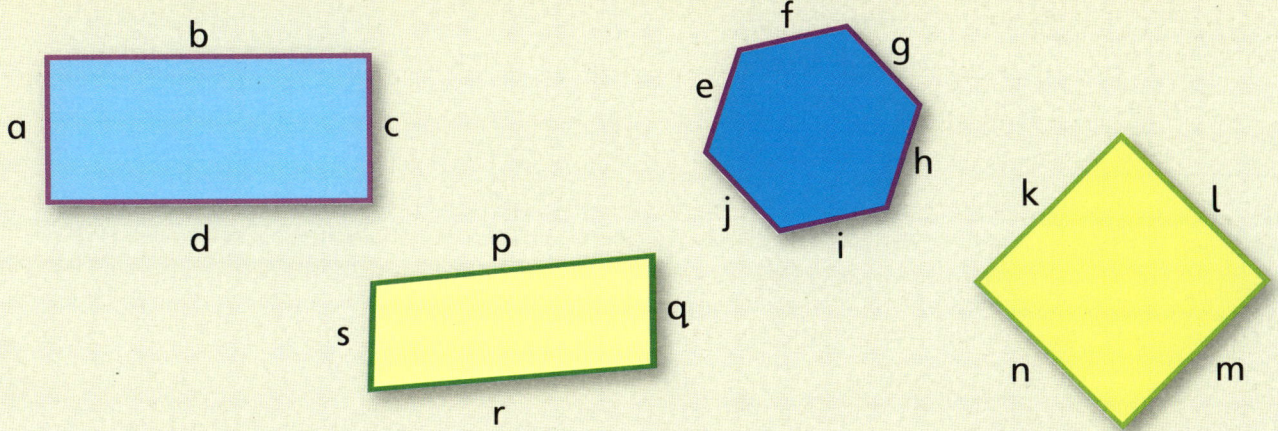

Write the letters of the sides that are:

1. parallel to b
2. perpendicular to m
3. parallel to r
4. perpendicular to a
5. parallel to f
6. parallel to c
7. perpendicular to k
8. parallel to h
9. parallel to s
10. parallel to j
11. parallel to d
12. parallel to q.

Draw a hexagon with only two parallel sides. Try to draw one with three parallel sides.

True or false?

13. All rectangles have two pairs of parallel sides.
14. A regular hexagon has three pairs of parallel sides.

Make up some true or false statements like these for your partner to answer.

I can identify parallel and perpendicular sides of shapes

Triangles

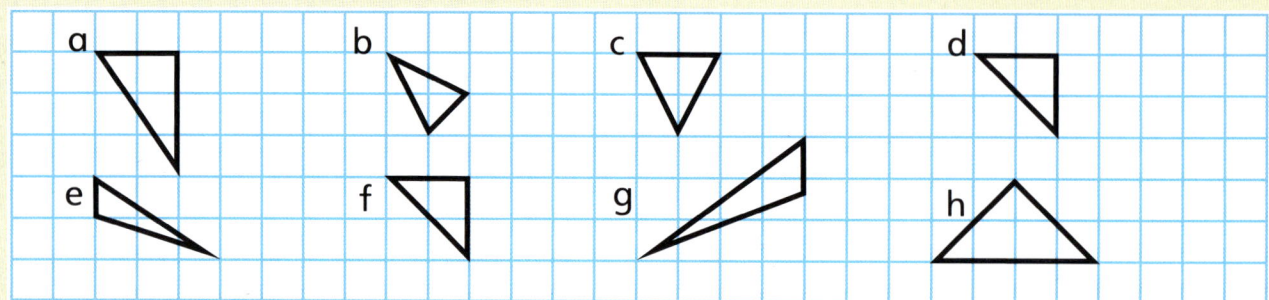

Which of these triangles have:

1. no equal sides?
2. two equal sides?
3. no equal angles?
4. two equal angles?
5. no right angles?
6. one right angle?

Which of these triangles are:

7. scalene?
8. not right-angled?
9. right-angled isosceles?
10. right-angled scalene?

11. Draw some different isosceles triangles on squared paper. Mark the equal sides.

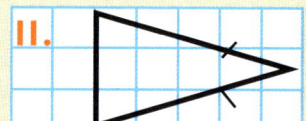

True or false?

12. A triangle cannot have more than one right angle.
13. A scalene triangle cannot be right-angled.
14. An isosceles triangle can be split into two equal right-angled triangles.
15. Right-angled triangles are never symmetrical.

I can describe the properties of triangles

Triangles

SPM2.6a

1 Write the name of each triangle.   1. (a) right-angled scalene

Choose from: isosceles   scalene   equilateral
right-angled scalene   right-angled isosceles

a 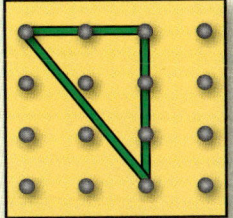   b 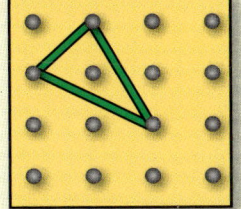   c 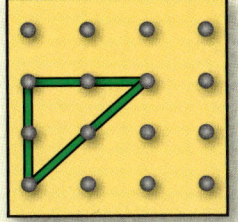   d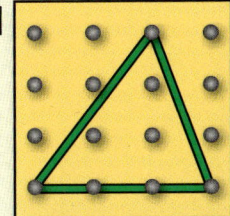

e 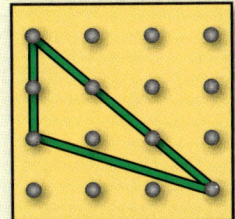   f 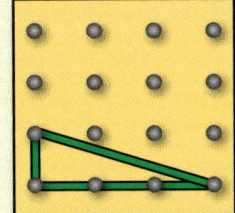   g 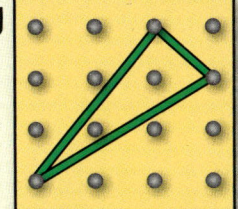   h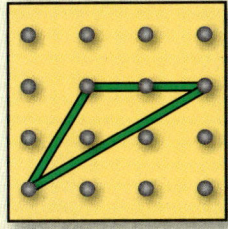

i 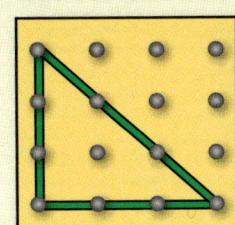   j 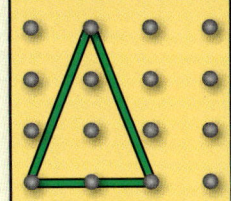   k 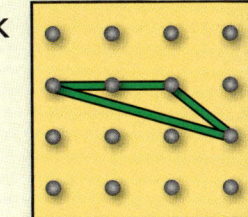   l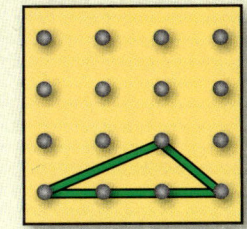

2 Which of these triangles are symmetrical?

Look at the triangles above. Draw a different symmetrical triangle on a similar dotted grid. Name it. Can you draw any more?

Which of the triangles above belongs in these areas?

3 blue   **4** yellow

5 pink   **6** green

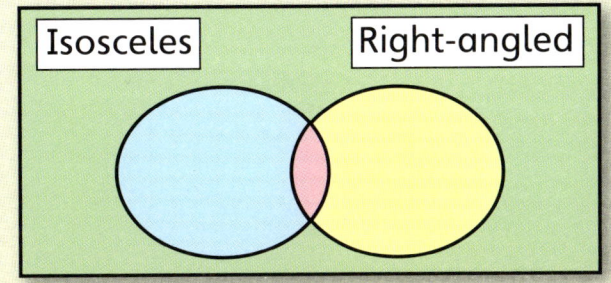

I can classify different triangles

27

Polygons

1 Are these polygons – yes or no?

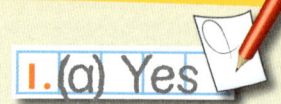

 a  b 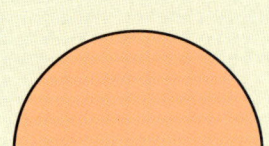 c  d

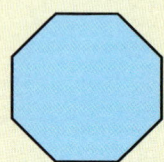 e  f 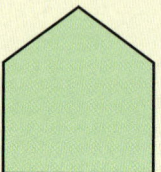 g 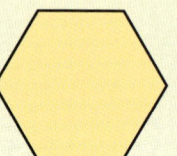 h

2 Which polygons are regular?

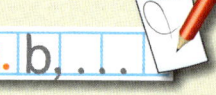

3 Which polygons are irregular?

4 Copy each shape, then draw its diagonals.

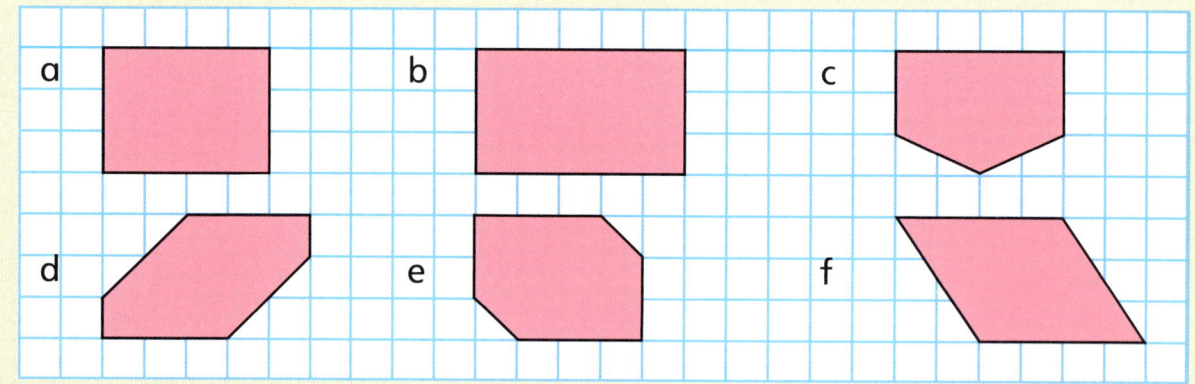

5 Write the name of each shape.

6 Write its number of diagonals.

I can describe regular and irregular polygons and understand their properties

Quadrilaterals

1. Copy and complete the table. For each statement and shape colour the box the correct colour.

🟩 always true   🟨 sometimes true   🟥 never true

	square	rectangle	parallelogram	rhombus	trapezium	kite	arrowhead
Has four sides							
Has all sides the same length							
Has one pair of opposite sides parallel							
Has two pairs of opposite sides parallel							
Has opposite sides equal							
Has adjacent sides equal							
Has line symmetry							
Has one or more right angles							
Has one or more obtuse angles							
Has one or more reflex angles							

Here are two quadrilaterals drawn on 3 × 3 dotted paper.

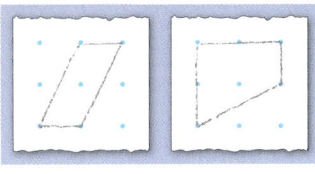

Draw your own quadrilaterals on dotted paper. How many different ones can you find?

You must decide whether these two count as the same.

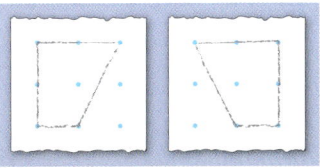

I can analyse the properties of quadrilaterals

Circles, ellipses and semi-circles

Read the descriptions below. They could describe more than one shape.

1. Which of them could describe a circle?
2. Which of them could describe a semi-circle?
3. Which of them could describe an ellipse?

a	b	c
I am drawn with just one curved line.	I am half a circle.	I have only two lines of symmetry.

d	e	f
I am drawn with just one line. The distance from my centre to my outline is always the same.	I have only one line of symmetry.	I have no straight sides.

g	h	i
I have one curved side and one straight side.	I have an infinite number of lines of symmetry.	I am drawn with just one line. The distance from my centre to my outline varies.

4. Write a full description of one of the three shapes so that your partner would recognise it. Do not use the word circle, semi-circle or ellipse in your description.

I can say some properties of circles, semi-circles and ellipses

Circles

Choose a word from the box to match each area or line.
Some words may be used twice.

| Circumference | Diameter | Radius | Chord |
| Circle | Semi-circle | Sector | Segment |

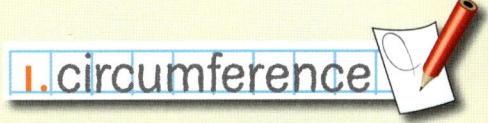

1. circumference

1   2   3   4

 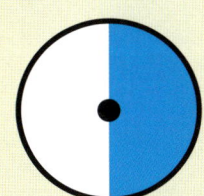 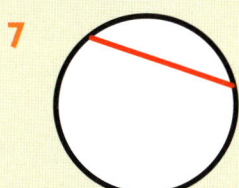 

5   6   7   8

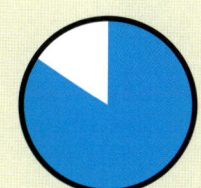   

9   10   11   12

Draw a circle and inside it draw two radii and a chord to make a triangle. What type of triangle is it? What different triangles can you make this way?

I can name the parts of a circle

SPM2.7a

3D objects

Find objects like the ones below. For each object:

a   Name the shape.
b   Write the number of vertices.
c   Write the number of edges.
d   Write the number of faces.

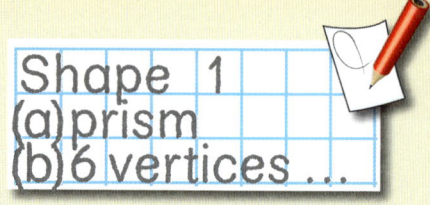
Shape 1
(a) prism
(b) 6 vertices …

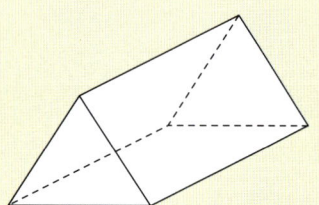
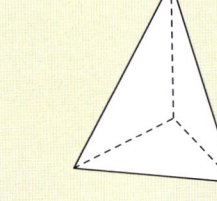
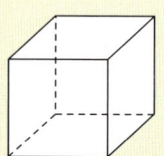
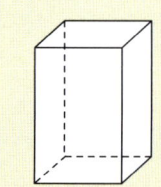
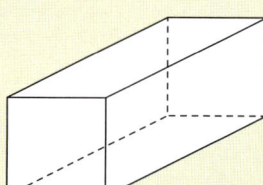
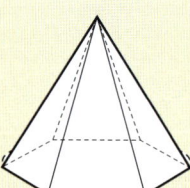
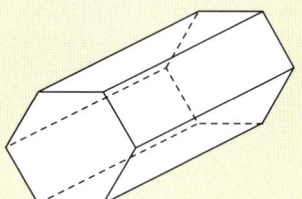
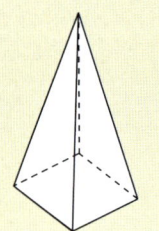
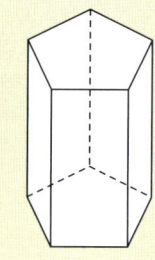
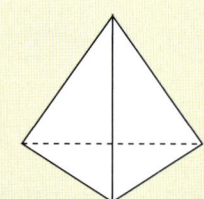
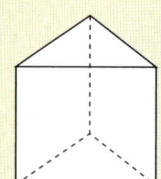
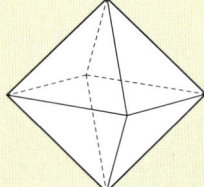

A mathematician called Euler discovered this fact: in polyhedrons, the number of faces + the number of vertices = the number of edges minus 2. Is this true for the objects above?

I can say the names and the numbers of faces, vertices and edges of 3D objects

32

3D objects

SPM2.7a

Write the name of each object. How many faces does it have?

1 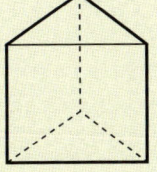   2 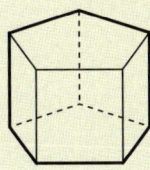   1. Cube, 6 faces

3 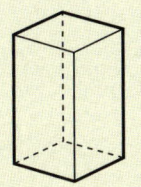   4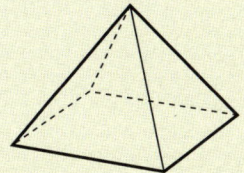

5 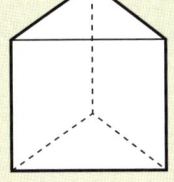   6 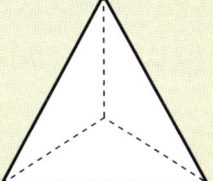   7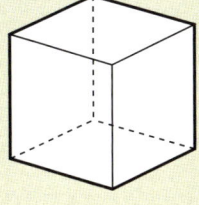

Draw all the faces of each object.

8 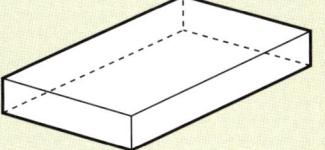   9 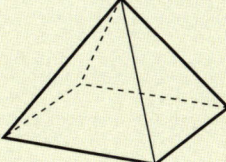   8.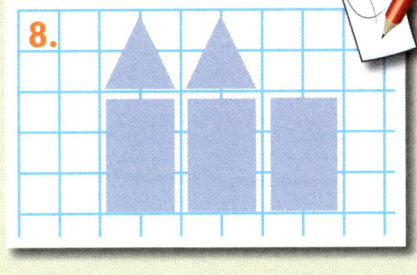

10    11    12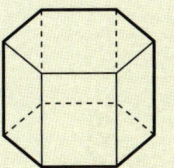

 Think of a 3D object with at least three pairs of parallel edges. Find examples of this object in the classroom. Try drawing it.

I can say the number and shapes of the faces of 3D objects

3D objects

Write the name of each object. Count the faces. Draw each different face on squared paper and name it. Say how many there are.

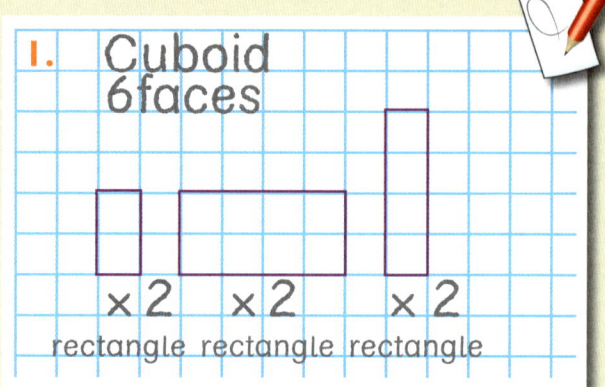

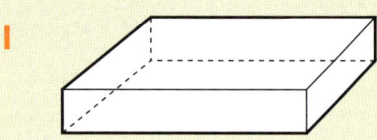 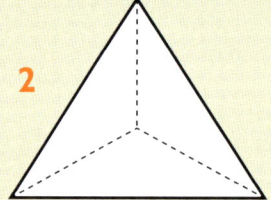 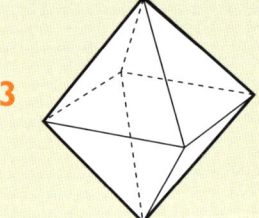

1   2   3

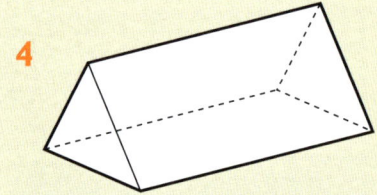 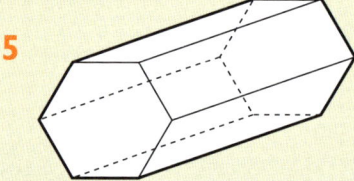 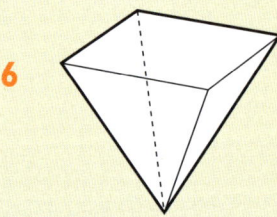

4   5   6

| hexagonal prism | octahedron | tetrahedron | square-based pyramid |

| cuboid | triangular prism |

Look at each object above. Does it have:

a   parallel faces?   b   perpendicular faces?

c   perpendicular edges?   d   parallel edges?

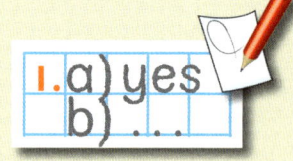

Can you think of (and draw) a 3D object with an odd number of faces and at least one pair of parallel edges?

Can you find more than one object like this?

I can describe the faces of 3D objects and say if they are parallel or perpendicular

Objects with curved faces

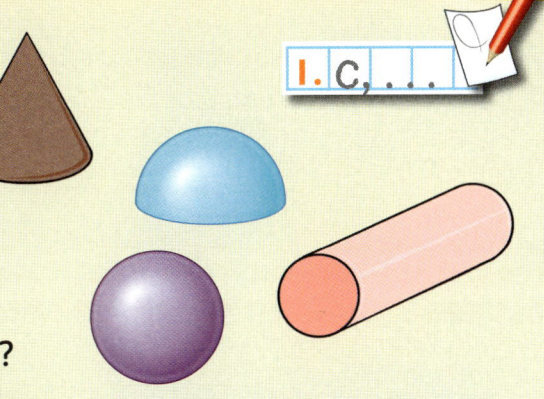

Read the descriptions below. There might be more than one answer for each one.

1. Which of these could describe a cone?
2. Which of these could describe a sphere?
3. Which of these could describe a cylinder?
4. Which of these could describe a hemisphere?

a I have two vertices.	b I have two flat faces.	c I have one vertex.
d I have three flat faces.	e I have one flat face.	f I have two edges.
g I have three edges.	h I have one edge.	i I have one flat face and a curved surface.
j I have no edges.	k I have no flat faces.	l I have no vertices.

5. Write a full description of one of these four shapes so that your partner would recognise it. Do not use the word cone, sphere, cylinder or hemisphere in your description.

I can say some properties of cones, spheres, hemispheres and cylinders

Nets of cones and cylinders

Match each net with the shape it makes.

1. Net 1 → Shape B

Net 1

Net 2

Net 3

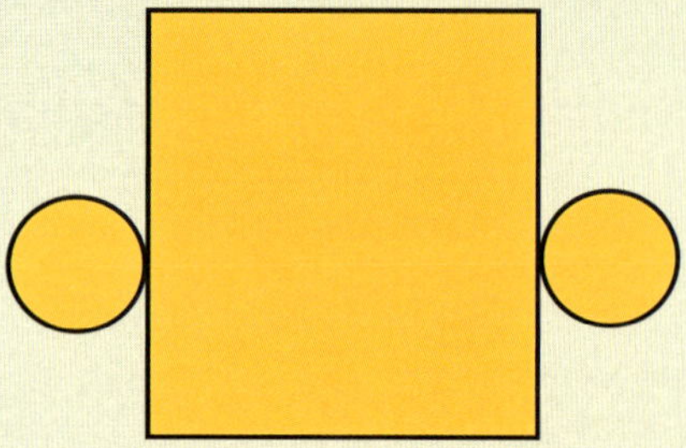

Net 4

Net 5

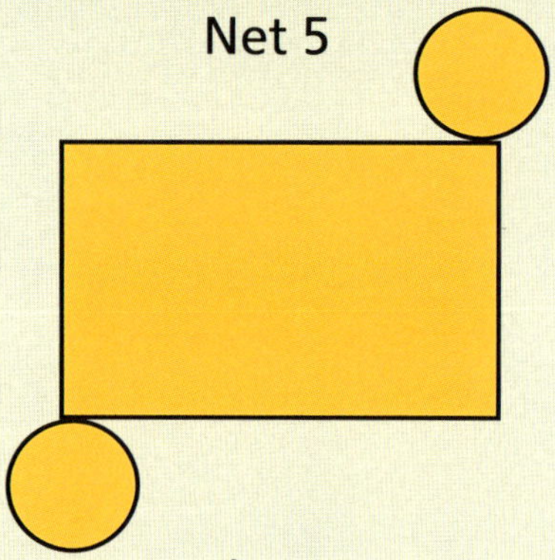

Net 6

Shape A

Shape B

Shape C

Shape D

Shape E

Shape F

Make your own nets to help you check your answers.

I can recognise nets of cones and cylinders

Combining shapes

James has three types of tiles. The sides of the shapes are all the same length, except for the longest side of the trapezium, which is twice as long as the other sides.

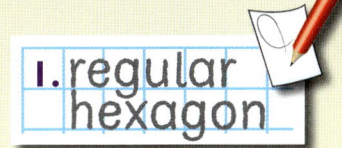

He combines tiles to make new shapes. Write the name of each new shape.

1    2 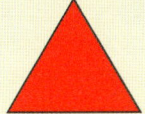   3

4    5    6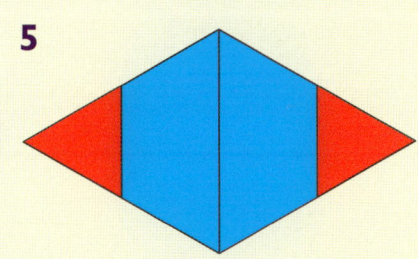

7   Use isometric dotted paper to find different ways to draw a regular hexagon using the trapezium, triangle and rhombus. Try to find at least six different ways.

8   Use isometric dotted paper to make some shapes of your own using the trapezium, triangle and rhombus. Name each new shape and the tiles used.

I can combine shapes to make other shapes

Tessellating

Follow these instructions to make your own tessellating shape.

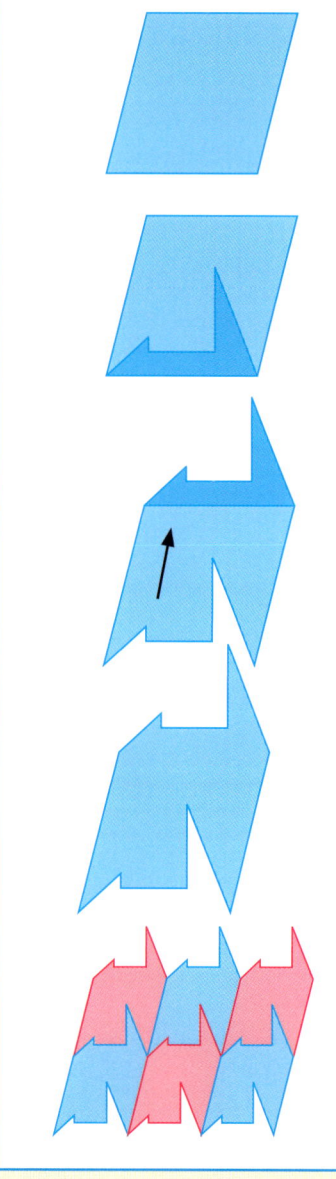

1 Start with any cardboard shape that tessellates.

2 Cut a shape from one side.

3 Slide the cut out piece to the opposite side.

4 Stick the pieces together.

5 Now draw round your new shape to make tiling patterns.

Try cutting a curved shape from one of the other sides and sliding it to the opposite side, like this:

I can make tessellating patterns

Angles in tessellations

These patterns have been made from tessellating single shapes or combinations of shapes.

Write the names of the shapes in each pattern, then work out the angle shown in yellow.
Remember what you know about angles about a point.

1. Regular hexagons
   yellow angle = 120°

1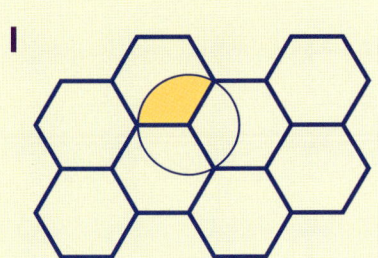

2

3

4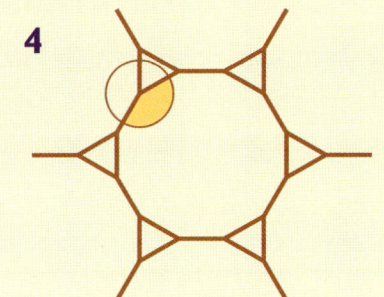

5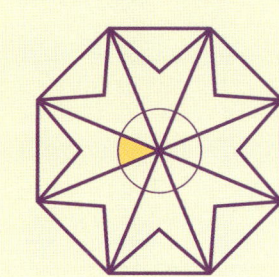

6

7

Talk to your partner about how you worked out each answer.

I can work out the size of angles in tessellating patterns

Symmetry

Copy and complete the pattern each time.
It must be symmetrical in both lines.

1 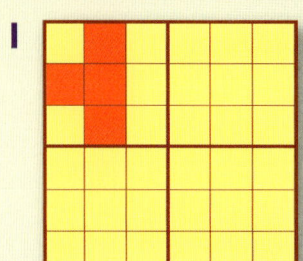 2 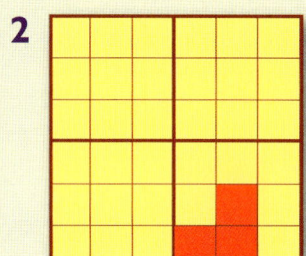 3 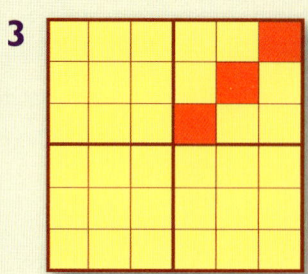 4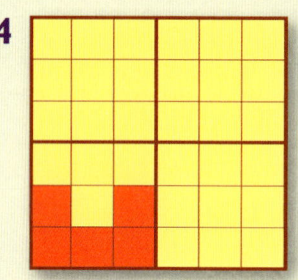

Write how many lines of symmetry each square has.

5 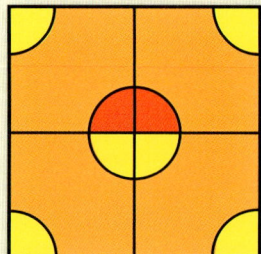 6  7  8

9 Show your partner where the lines of symmetry are on patterns 5 to 8. Do they agree?

Use squared paper.
Draw two lines on one square.

Explore rotating the square or reflecting it to make a pattern.

Does reflecting it produce the same pattern as rotating it?

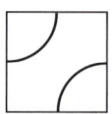 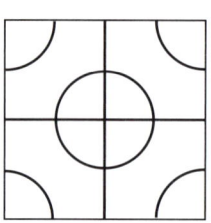

I can make symmetrical patterns

Reflections

Use squared paper.
Draw each shape and the axes.
Draw the shape's reflection across both axes.

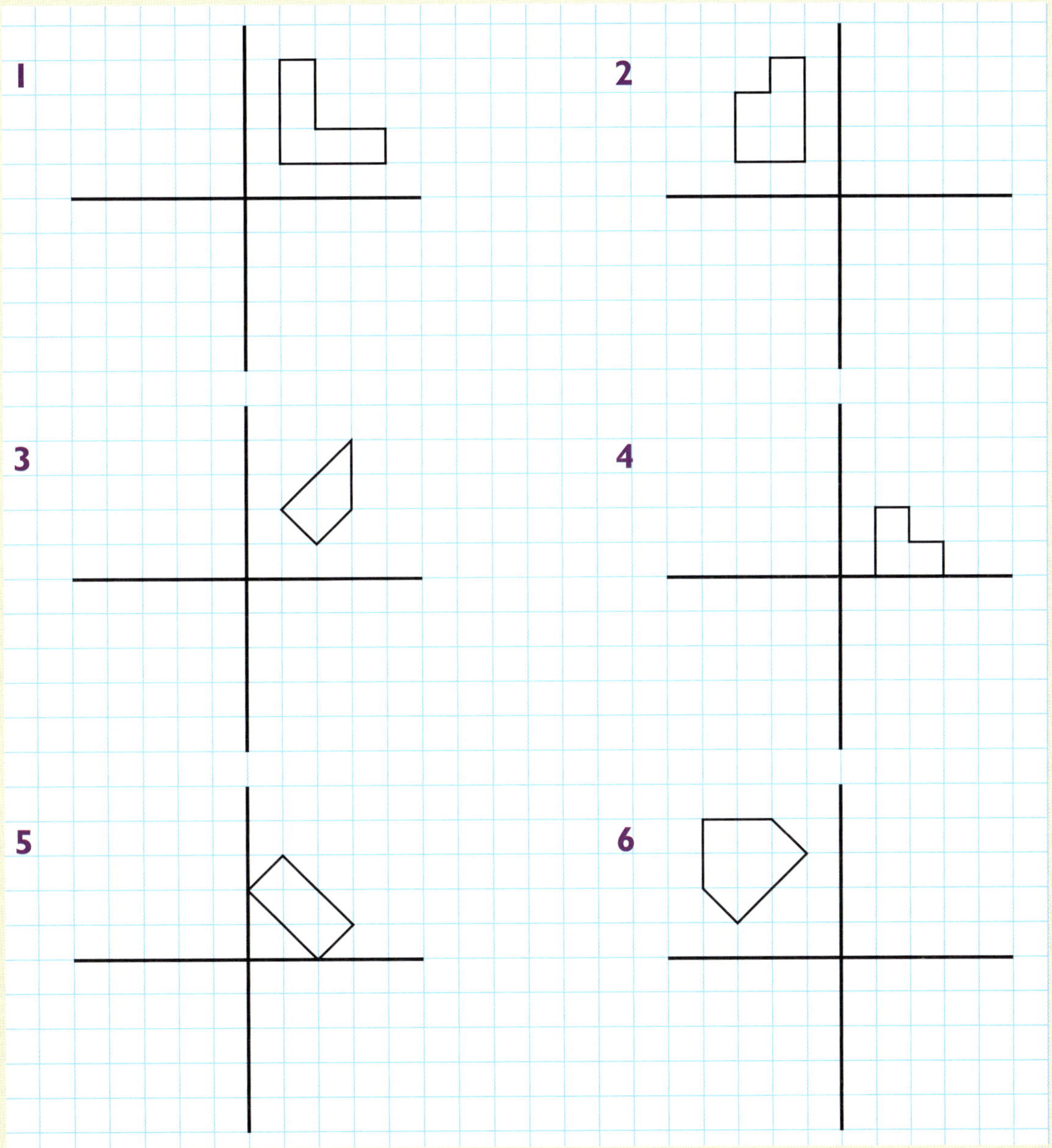

1
2
3
4
5
6

Reflect different quadrilaterals in two mirror lines.
Try quadrilaterals with no lines of symmetry.

I can reflect shapes in two mirror lines

41

Rotations

Draw the position after a rotation of 90° clockwise about the point.

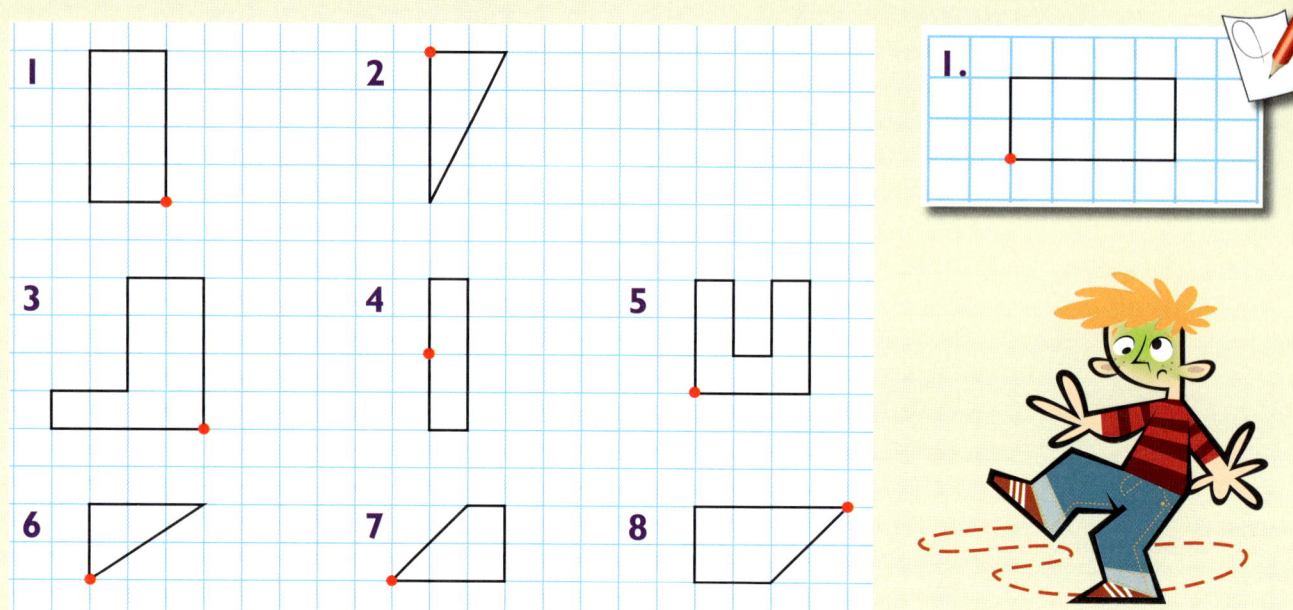

Find the matching shape after a rotation of 90° clockwise about the point.

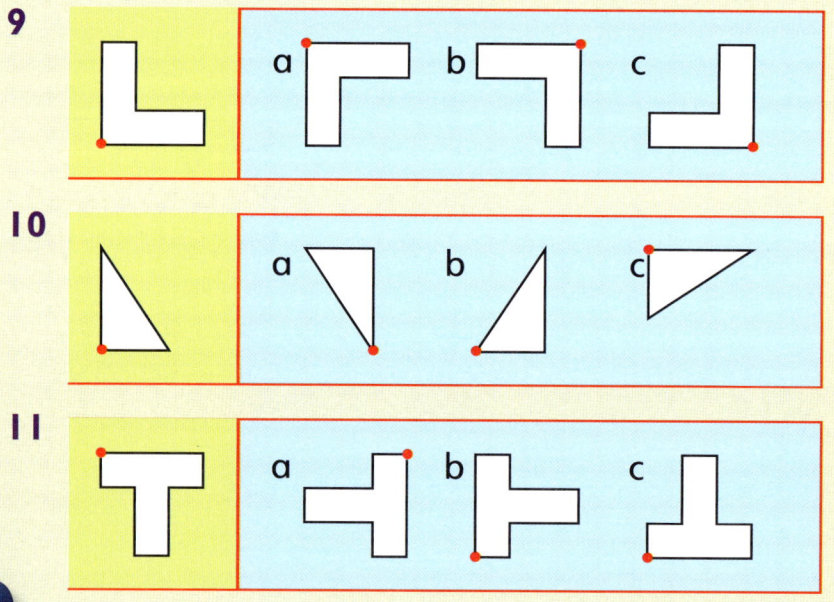

Explore rotating the first letter of your name clockwise through 90°, 180°, 270°.

I can rotate shapes through 90° about different points

Rotations

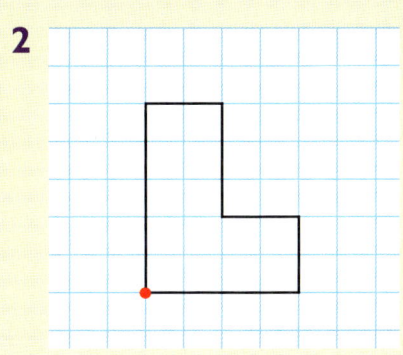

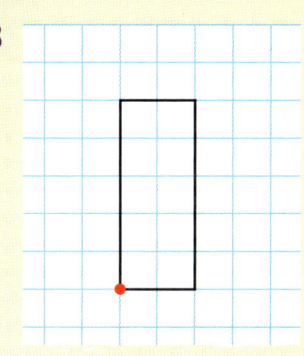

Rotate each shape about the point through:

a 180° clockwise   b 90° anticlockwise   c 270° clockwise

d 180° anticlockwise

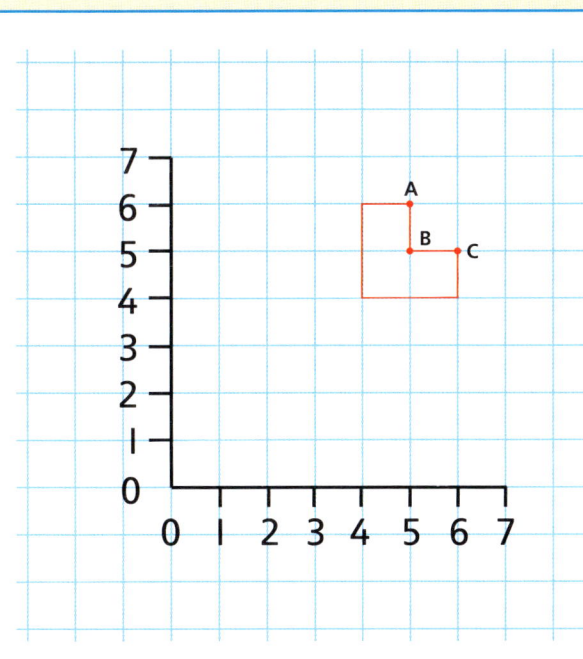

Rotate this shape 90° clockwise about the point (4, 4).

Write the new coordinates of points A, B, C.

Explore rotating about 180°, 270°.

I can rotate shapes through 90°, 180° and 270°

Patterns

Choose a tile. Fill a 2 × 2 grid by:

a   translating (sliding) it   b   rotating it   c   reflecting it

translation   rotation   reflection

Repeat for each tile and each type of transformation.

Try these rectangular blocks.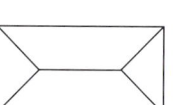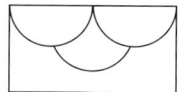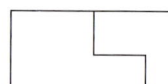

Rotate, reflect or slide them.

What patterns can you create?

Colour some to create a spectacular tiling pattern.

I can make translations, rotations and reflections

Polygons

Copy each shape onto dotted paper, then draw its diagonals.

1 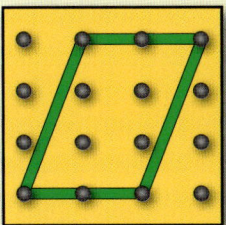 2 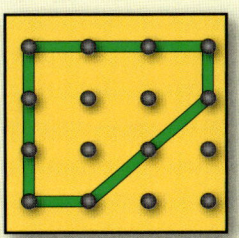 3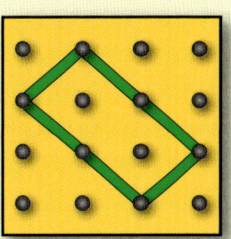

Write the name of each shape, and its number of diagonals.

4 Draw a pentagon and a hexagon on dotted paper. Make them fairly big. Find how many diagonals each shape has.

 Do all hexagons have the same number of diagonals? And if so, why?

True or false?

5 The sides of a regular polygon are all equal.

6 A regular quadrilateral is called a square.

7 A triangle never has a diagonal.

8 All pentagons have a total of five diagonals.

9 The diagonals of a square are perpendicular to each other.

10 Regular polygons are always symmetrical.

11 A regular triangle is called an equilateral triangle.

12 A rectangle is an irregular polygon.

13 If all the diagonals of a polygon are the same length, it must be a regular polygon.

I can solve problems relating to the diagonals of polygons

Triangles

1. Copy the diagrams onto isometric paper.

   How many equilateral triangles can you find in each diagram?

   Write how many triangles there are of each size.

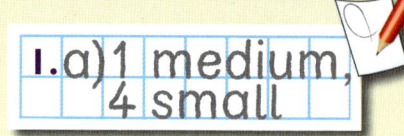
1.a) 1 medium, 4 small

a

b

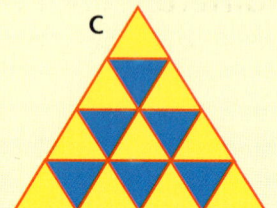
c

Draw the next diagram in the series. How many triangles here?

Draw a large square, then draw its diagonals.

Cut it out, then cut along the diagonals to make four triangles.

Find and draw triangles you can make by combining:
- two triangles
- three triangles
- four triangles.

Name each triangle.

Investigate other shapes you can make.

Name them and draw their lines of symmetry.

I can investigate and solve problems about triangles

Quadrilaterals

Use dotted paper to make a parallelogram within a 4 × 4 square. Here are two. How many different ones can you make altogether?

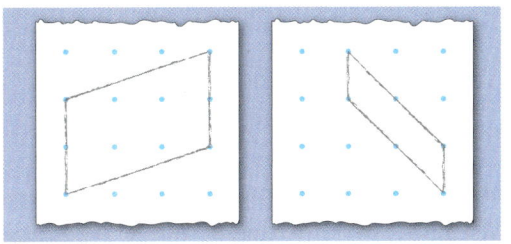

True or false?

1. A rhombus is a parallelogram with four equal sides.
2. A trapezium has one pair of parallel sides.
3. A square is a type of rhombus.
4. A rectangle is a type of parallelogram.
5. A parallelogram has two pairs of equal sides.
6. A parallelogram always has line symmetry.
7. A trapezium can have a right angle.
8. A parallelogram can have two obtuse angles.
9. The angles of a rhombus are never equal.
10. A trapezium is always symmetrical.
11. The diagonals of a parallelogram can be the same length.
12. If one angle of a rhombus is 90°, it must be a square.

Draw this rhombus on isometric paper.

Draw some others.

Draw the diagonals of each rhombus.

What do you notice about where the diagonals cross?

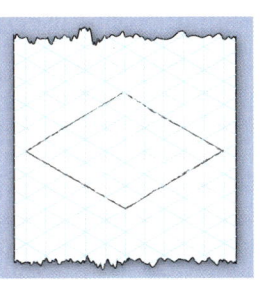

I can investigate and describe quadrilaterals and create them

Author Team: Lynda Keith, Hilary Koll and Steve Mills
Consultant: Siobhán O'Doherty

Published by Pearson Education Limited, a company incorporated in England and Wales, having its registered office at Edinburgh Gate, Harlow, Essex, CM20 2JE. Registered company number: 872828

www.pearsonschools.co.uk

Text © Pearson Education Limited 2012

First published 2012

15 14 13 12
10 9 8 7 6 5 4 3 2 1

British Library Cataloguing in Publication Data
A catalogue record for this book is available from the British Library

ISBN 978 0 4350 7744 0

Copyright notice
All rights reserved. No part of this publication may be reproduced in any form or by any means (including photocopying or storing it in any medium by electronic means and whether or not transiently or incidentally to some other use of this publication) without the written permission of the copyright owner, except in accordance with the provisions of the Copyright, Designs and Patents Act 1988 or under the terms of a licence issued by the Copyright Licensing Agency, Saffron House, 6–10 Kirby Street, London EC1N 8TS (www.cla.co.uk). Applications for the copyright owner's written permission should be addressed to the publisher.

Typeset by Debbie Oatley @ room9design and revised by Mike Brain Graphic Design Limited, Oxford
Illustrations © Harcourt Education Limited 2006–2007, Pearson Education Limited 2011
Illustrated by John Haslam, Anthony Rule, Piers Baker, Matt Buckley, Andrew Hennessey, Eric Smith, Gary Swift, Nigel Kitching, Mark Ruffle, Q2A Media, Jonathan Edwards, Stephen Elford, Sim Marriott, Andrew Painter, Fred Blunt, Emma Brownjohn, Tom Percival, Tom Cole, Debbie Oatley
Cover design by Pearson Education Limited
Cover illustration by Volker Beisler © Pearson Education Limited
Printed in the UK by Scotprint

Acknowledgements
The Publishers would like to thank the following for their help and advice:
Liam Monaghan
Hilary Keane
Stephen Walls

Every effort has been made to contact copyright holders of material reproduced in this book.
Any omissions will be rectified in subsequent printings if notice is given to the publishers.